Introduction

Interactive physics has been written to provide a book that is designed around the kind of questions and explanations that the exams require. It should be used in conjunction with the "E our tutor will provide you with. This contains hundreds of exam style questions fr   for your exams. Many books include these questions in the book itself, but with the   to refer to it again and again, this booklet is separate.

The book initiates you to the need to use SI units for all your work.   that require you to change between units, but understanding them is essential for success in physics.   ng the composite unit for a physics quantity will give you a hint as to the equation that you need to use to calculate it.

Important equations are left inside boxes, with the physical quantities listed to the right, together with the relevant base units that should be used in conjunction with it. Failing to convert to these units before using the equation will guarantee that the answer will be WRONG!

Each chapter has a section of revision questions to try out your new-found knowledge. Work through these in a separate workbook and hand the answers to your tutor for marking.

The book grew out of my 1-2-1 tutoring with several students over the recent years. It is designed to work in small, simple, units that can be taught in one, one-hour tutorial session, with homework for the student to complete between session.

It would be tempting to ignore the sections that do not apply to your exam board, but I suggest that you work through the whole book. Physics is a fascinating subject and worth learning for its own sake, but also because, if you choose to study A-level physics, the parts of the book your syllabus did not cover will often pop up.

The end of the book has a list of all the relevant equations for your syllabus, and ones from the other exam boards.

Acknowledgements

Where I have used art and images from the internet, I have tried to use only those in the "public domain" or those that have a notice that the image may be freely used. However, it may be possible that some sites have not had explicit notices in respect of this, or where the site has used images from other sites without acknowledging the source.

Contents

Chapter 1. Measurement and Units 7
  1.1 Units in physics 7
  1.2 Scientific notation 8
  1.3 Measurement 9
    Measurement of Mass 9
    Measurement of Time 10
    Measurement of Length 11
    Methods of measurement 12
    Parallax errors 13
  1.4 Questions: 13

Chapter 2. Forces and their effects 15
  2.1 Forces 15
    Types of force 15
    What can forces do? 15
  2.2 Weight as a force 16
    Stretching a spring 16
    Force constant 18
    Measuring the force constant of a spring 19
  2.3 Questions 20

Chapter 3. Mass and weight 21
  3.1 Differences between mass and weight 21
    The gravitational field strength 22
  3.2 Vectors and Scalars 23
    Adding vectors and scalars 23
    Adding Vectors Which Are Not In Line 24
  3.3 Questions: 25
  3.4 Forces can change the motion of an object 26
  3.5 Newton's First Law 26
  3.6 Newton's Second Law 27
  3.7 Questions 29
  3.8 Newton's Third Law 29
  3.9 Questions 30

Chapter 4. Speed, velocity and acceleration 31
  4.1 Acceleration 32
  4.2 Questions 32
  4.3 Using graphs to describe motion 33
    Distance-time graphs (plotting how far something has moved against the time taken) 33
    Displacement 34
    Velocity-time graphs 35
  4.4 Stopping distances 37
  4.5 Questions 38
  4.6 Circular Motion 39
  4.7 Mechanics - SUVAT 41
    Worked example 45
    Worked example 46
  4.8 Questions 48

Chapter 5. Turning Forces - Moments 49
  Examples of Moments in action 49

Chapter 6. Moments in balance 51
  Worked example 51
  6.2 Questions 52

Chapter 7. Centre of mass and equilibrium 54
  7.1 Pendulums 59
  7.2 Questions 61

Chapter 8. Density 62
  8.1 Measuring density 62
    Worked example 63
  8.2 Questions 63

Chapter 9. Pressure 64
  Practical examples 64
  9.2 Questions 65
  9.3 Pressure in liquids 66
    Calculating pressure in liquids 67
  9.4 Questions 69
  9.5 Hydraulics 69
    Liquids do not compress 69
    When the area of the parts of the pump change 69
  9.6 Atmospheric pressure 71
    Atmospheric Pressure 72
    Measuring Atmospheric Pressure 72
    Measuring Pressure Difference 73
  9.7 Questions 74

Chapter 10. Energy 75
  10.1 Types of energy 75
  10.2 Transformation of energy. 75
    Principle (or law) of Conservation of Energy 75
    Example 76
  10.3 Questions: 76
  10.4 Energy transfer diagrams (or Sankey diagrams) 77
    Efficiency 78
  10.5 Questions 78
  10.6 Calculating Potential, Elastic and Kinetic Energy .. 79
    Calculating gravitational potential energy (G.P.E.) . 80
    Calculating stored elastic potential energy 80
  10.7 Calculating kinetic energy (K.E.) 81
  10.8 Questions 81
  10.9 K.E. and G.P.E Problems 83
    Worked example 1 – calculating loss of E_p 83

Table of Contents

- Worked example 2 – calculating resulting velocity 83
- 10.10 Questions: 84
- Worked example 3 – calculating height 85
- 10.11 Questions 85
- Chapter 11. Energy resources 86
  - 11.1 Electricity generation 86
  - 11.2 Non-Renewable fuels 87
  - 11.3 Electricity from water 88
    - Tidal power 88
    - Hydroelectric power 89
  - 11.4 Electricity from air 90
    - Wind farms and Areogenerators 90
  - 11.5 Other schemes 90
  - 11.6 Renewable energy resources 92
- Chapter 12. Energy, work and power 94
  - 12.1 Work 94
  - 12.2 Questions 94
  - 12.3 Power 94
    - Worked example 95
  - 12.4 Electric power 96
    - Worked example 96
  - 12.5 Questions 96
- Chapter 13. Momentum 97
  - 13.1 Collisions, Impulse 97
  - 13.2 Conservation of Momentum 98
  - 13.3 Questions 99
- Chapter 14. Astronomy 100
  - 14.1 The Big Bang and Age of the Universe 100
    - Formation of stars and galaxies 100
  - 14.2 The Universe 103
  - 14.3 The life of a star 103
  - 14.4 Knowing these things 106
  - 14.5 Questions: 107
  - 14.6 Satellites 108
    - Orbit speeds of satellites 109
  - 14.7 The Solar System 110
  - 14.8 Planets 111
    - Mercury (0.4AU) 112
    - Venus (0.7AU) 112
    - Earth (1AU) 112
    - Mars (1.5AU) 113
    - Asteroid Belt and Dwarf Planets 113
    - Jupiter (5.0AU) 113
    - Saturn (9.5AU) 114
    - Uranus (19AU) 114
    - Neptune (30AU) 115
  - 14.9 Orbiting times 115
    - In between the gas giants 116
    - The solar system beyond Neptune – The Kuiper Belt and the Oort cloud 116
    - Comets 117
  - 14.10 Questions 118
- Chapter 15. Astrophysics 119
  - 15.1 Colour 119
  - 15.2 Brightness 120
  - 15.3 Hertzsprung-Russell diagram 120
  - 15.4 Blackbody radiation 122
- Chapter 16. Cosmology 123
  - 16.1 The Hubble Law and Doppler Shift 123
  - 16.2 Evidence that demanded the Big Bang theory 124
  - 16.3 Dark Energy and dark matter 125
- Chapter 17. The kinetic model 126
  - 17.1 Properties of the states of matter 126
  - 17.2 Evidence for moving particles 127
  - 17.3 Movement and energy of particles 128
    - The Kelvin Temperature Scale 129
  - 17.4 Boiling and Evaporation 129
    - Evaporation 129
    - Cooling effect of evaporation 130
  - 17.5 Questions 131
  - 17.6 Thermal expansion 133
  - 17.7 Expansion in Solids 133
    - Using expansion 133
    - Problems and solutions to problems with expansion 135
  - 17.8 Expansion of liquids 136
  - 17.9 Questions 137
- Chapter 18. Measurement of Temperature 139
  - 18.1 Temperature 139
  - 18.2 Liquid in Glass Thermometers 139
    - Key Terms used in thermometers 140
  - 18.3 The Thermocouple 140
  - 18.4 Questions 141
- Chapter 19. The Gas Laws 142
  - 19.1 Boyle's Law (Constant Temperature) 142
  - 19.2 Worked example 143
  - 19.3 Worked Example 143
  - 19.4 Questions 143
  - 19.5 Gay-Lussacs' Law (aka the Pressure-Temperature Law) 144
- Chapter 20. Specific Heat Capacity 146

- 20.1 Worked example 146
- 20.2 Thermal Capacity 146
- 20.3 Power and Energy 147
  - Experiment to determine the specific heat capacity of water 147
  - Finding other unknown entities in the specific heat capacity equation 148
- 20.4 Worked example 149
- 20.5 Questions 149

Chapter 21. Latent Heat 150
- 21.1 Specific latent heat of fusion (or freezing) .. 150
- 21.2 Specific Latent Heat of Vaporization 152
  - Performing experiments to determine specific heats 152
- 21.3 Questions 154

Chapter 22. Heat Transfer 156
- 22.1 Conduction 156
  - Demonstrations of conduction 157
- 22.2 Convection 158
  - Examples of convection in action 159
- 22.3 Radiation 161
  - Emitters and absorbers 162
- 22.4 Questions 165

Chapter 23. Common Wave Terminology 167
- 23.1 Background 167
- 23.2 Wave Terminology 167
- 23.3 Time Period 168
- 23.4 The Wave Equation 169
- 23.5 Types of waves 169
  - Longitudinal waves 169
  - Longitudinal Waves 170
- 23.6 Normals 171

Chapter 24. Wave Effects 173
- 24.1 Ripple Tanks 173
- 24.2 Reflection 173
- 24.3 Refraction 175
- 24.4 Diffraction 177
- 24.5 Questions 178

Chapter 25. Light 181
- 25.1 Properties of light 181
- 25.2 Reflection 181
- 25.3 Reflection of light on a plane mirror 182
  - Reflection of an image in a plane mirror 183
- 25.4 Questions 185
- 25.5 Refraction 186
- 25.6 Calculating refraction indexes for a single substance 190
  - Worked example 191
  - Worked example 191
- 25.7 Questions 191

Chapter 26. Total Internal Reflection (TIR) 193
- 26.1 Calculating the critical angle 194
- 26.2 Uses of Total Internal Reflection 196
- 26.3 Optical Fibres 197
- 26.4 Questions 198

Chapter 27. Lenses 199
- 27.1 Convex Lenses 199
- 27.2 Concave Lenses 199
- 27.3 Real and virtual images 200
- 27.4 How images formed by lenses 200
- 27.5 Dioptre – Lens Power 202
- 27.6 Worked ray tracing example 202
- 27.7 Questions 203

Chapter 28. Dispersion of Light 204

Chapter 29. The Human eye 205
- 29.1 Parts of the eye 205
  - Cornea 205
  - Pupil 205
  - Iris 205
  - The Eye lens 206
  - Ciliary muscles and suspensory ligaments 206
  - Retina 206
- 29.2 Forming the image 206
  - Short-sightedness (myopia) 206
  - Long sightedness (hyperopia) 207
  - Correcting errors in the shape of the eye 207
- 29.3 Comparing the eye with a digital camera ... 209

Chapter 30. The Electromagnetic Spectrum 210
- Gamma Rays 211
- X-Rays 211
- Ultraviolet 212
- Visible light 212
- Infrared 212
- Microwaves 212
- TV and Radio 212
- Questions 212

Chapter 31. Sound waves 213
- Experiment to show that air molecules are required to transmit sound waves 214
- Speed of sound and echoes 214
- Reflection of sound and echoes 214
- 31.2 Worked examples 214

- 31.3 Terminology of sound waves 215
- 31.4 Oscilloscopes tracing sound waves 215
- 31.5 Questions 216

Chapter 32. The Atom and the Electron 217
- 32.1 Atomic Structure 217
- 32.2 Electrons 218

Chapter 33. Magnets and Magnetism 219
- 33.1 How the electron is involved 219
- 33.2 Properties of magnets 219
  - Magnetic flux lines 220
  - Ferromagnetics 220
  - Hard and Soft Magnetic Materials 220
- 33.3 Magnetic induction – making magnets 220
  - Stroking Method 221
  - Electrical Method 221
- 33.4 Destroying magnetism – demagnetisation .. 222
  - Hammering 222
  - Heating 222
  - Coil of wire using A.C. current 222
- 33.5 Magnetic field 222
  - Observing magnetic field lines 223
  - Observing the direction of the field lines 223
  - When magnets attract 223
  - When magnets repel 224
  - The Earth's magnetic field 225
  - Plotting the magnetic field 225
- 33.6 Magnetic effect of a current 226
- 33.7 The coil 227
- 33.8 Electromagnets 228
  - Uses of electromagnets 229

Chapter 34. Electrostatics 233
- 34.1 Electrostatic materials 233
  - Attraction and repulsion 234
- 34.2 Conductors and insulators 234
- 34.3 Electric Fields 235
- 34.4 Attraction of uncharged objects 236
  - Earthing 237
- 34.5 The Van De Graaf Generator 238
- 34.6 Charging by induction 239
- 34.7 Unit of charge 240
- 34.8 Uses of electrostatic charge 241
- Questions 243

Chapter 35. Electricity 244
- 35.1 Current 244
  - Charge and current 244
  - Current direction 245
- 35.2 Questions: 245
- 35.3 Voltage 246
- 35.4 Questions: 247

Chapter 36. Types Circuits 249
- 36.1 Series circuits 249
- 36.2 Parallel circuits 250
- 36.3 Questions: 252

Chapter 37. Resistance and Ohm's law 254
- 37.1 Resistance 254
- 37.2 The heating effect of resistance 255
- 37.3 Different types of resistors 256
- 37.4 Questions 256
- 37.5 Ohm's law 256
- 37.6 Current and PD graphs 258
- 37.7 Questions: 261

Chapter 38. Resistors in circuits 262
- 38.1 When placed in Series 262
- 38.2 When placed in parallel 263
  - Same Resistance 263
  - Different resistances 263
  - Examples 264
  - When combining series and parallel sets of resistors: 264
- 38.3 Questions: 266

Chapter 39. Electrical energy 268
- 39.1 Energy transferred 268
  - Worked Example 269
  - Worked Example 269
- 39.2 Electrical power 270
  - Worked example 271
- 39.3 Electrical power in circuits 271
- 39.4 Questions 271
- 39.5 Power in resistors 271
  - Worked examples 272
- 39.6 Questions 272

Chapter 40. Electrical safety 274
- 40.1 Circuit Breakers 274
- 40.2 Fuses 274
- 40.3 Fused plugs 275
- 40.4 Safety features 275
- 40.5 Hazards 276
- 40.6 Calculations used in electrical safety 276
  - Worked Example 276
- 40.7 Earthing 277
- 40.8 Double insulation 277
- 40.9 Dealing with electrical accidents 278

Chapter 41. Effects of electromagnetism 280
- 41.1 The generator 280
- 41.2 How to find the direction of the induced current 281
  - Fleming's Right-hand Rule 281
- 41.3 Force on a conductor in a magnetic field 282
- 41.4 Induced Current in a Coil 282
- 41.5 Observations 284
- 41.6 Faraday's Law 285
- 41.7 Lenz's Law 285
  - Finding the direction of the induced current in the coil 287
  - Typical uses of induced currents 287
- 41.8 Questions 287
- 41.9 The Motor effect 289
- 41.10 Fleming's Left Hand rule 290
  - Turning effect in a coil 291
- 41.11 Questions 291
- 41.12 Making use of electromagnetic induction. 292
  - The AC generators (alternator) 293
  - The DC motor 294
  - Observations 296
- 41.13 Questions 296

Chapter 42. Transformers 298
- Transformer Equation 300
- Worked example: 300
- 42.2 Questions 301
- Power through a transformer 301
- Worked example 302
- Transformer efficiency 303
- Worked Example 303
- 42.3 Power across the country – the national grid system 304
  - Local distribution of power 305
  - Long distance transmission of power 305
- 42.4 Questions 306

Chapter 43. Electronics 308
- 43.1 Electronic control systems 308
  - Control Systems 308
- 43.2 Main circuit components 309
  - Diodes 309
  - Transistors 312
- 43.3 Questions 314
  - How does it work? 316
  - Capacitor for a time-delay switch 316
- The thermistor 317
- The light-emitting diode (LED) 318
- 43.4 Questions 319

Chapter 44. Digital electronics 321
- 44.1 Analogue 321
- 44.2 Digital 321

Chapter 45. Logic gates 323
- 45.1 Truth tables 323
- 45.2 AND logic 323
- 45.3 OR Logic 324
- 45.4 NOT Gates (called an "inverter") 325
- 45.5 NAND Gates 326
- 45.6 NOR Gates 327
- 45.7 Combining logic gates 328
- 45.8 Questions 329

Chapter 46. Cathode ray oscilloscopes (CRO) 331
- 46.1 The CRO Display and controls 333
  - The display 334
- 46.2 Studying waveforms 335
- 46.3 Measuring voltage 335
- 46.4 Questions 337

Chapter 47. Atomic Physics 338
- 47.1 The Atom 338
  - Atomic Model 338

Chapter 48. The Model of the atom 342
- Nomenclature 343
- Questions 346

Chapter 49. Radioactivity 347
- 49.1 Background Radiation 347
- 49.2 Activity 347
- 49.3 Forms of Radioactivity 347
- 49.4 Alpha (α) Particles 348
- 49.5 Beta (β) particles 348
- 49.6 Gamma (γ) rays (not particles) 349
- 49.7 Summary 350
- 49.8 Deflection 350
- 49.9 Questions 352
- 49.10 Ionisation 353

Chapter 50. Detection of radiation 354
- 50.1 Photographic film 354
- 50.2 Gold leaf electroscope 354
- 50.3 Geiger-Muller Tube 354
- 50.4 Cloud Chamber 355
- 50.5 Spark Counter 356

Chapter 51. Uses of radioactivity 357
- 51.1 Tracers 357
- 51.2 Radio therapy 357

- 51.3 Thickness measurement (guaging) 357
- 51.4 Carbon 14 Dating 357
- 51.5 Sterilising 358
- 51.6 Question 358
- 51.7 Safety precautions when handling radioactive materials 359

Chapter 52. Radioactive Decay 360
- 52.1 Decay balancing equations 360
  - Alpha Decay 360
  - Beta Decay 360
  - Gamma Decay 361

Chapter 53. Half Life 362
  - Worked Example 363
  - Worked Example 364
- 53.2 Questions 364

Chapter 54. Fission 365
- 54.1 Chain reactions 366
  - Uncontrolled nuclear chain reactions 366
  - Controlled nuclear chain reactions (nuclear power stations) 367
  - Issues involved in nuclear power 369
- 54.2 Fusion 370
- 54.3 Questions 372

List of Figures 374
Index   382

Chapter 1. Measurement and Units

1.1 Units in physics

SI (SYSTEME INTERNATIONAL D'UNITES) UNITS are used throughout all science and physics especially. Understanding these is essential. The following table lays out the main SI units that you need to know – these are known as Base Units:

Table 1 - The Seven SI Base Units

Physical Quantity	Name of unit	Symbol for Unit	Example
Mass	Kilogram	Kg	80Kg (typical mass for a human male)
Time	Seconds	S	1s (the time it takes to say "elephant")
Length	Metre	m	1.8m (typical height for a human male)
Current	Ampere	A	13A is the typical current going through a washing machine
Temperature	Kelvin	K	300K is a fairly normal temperature in the UK
Amount of substance	Mole	mol	This is covered in Chemistry
Luminous intensity	Candela	cd	This is not covered until University

You need to become familiar with the first three (the ones that are not shaded). You might have to convert to the base units from others before you can use them in any equations, such as from feet to metres or miles to metres for length. In an exam you would typically be provided with a way of performing this conversion.

Most of the units are described to make sense at the scale of being a human, so 1s makes sense to us as the time it takes to say a word or 1m it fairly close to the height of a (short) person. They do not make a lot of sense when we think in sub-atomic scales or scales of the Universe, and so, as physicists are somewhat lazy, we have developed ways of making this easier using prefixes:

Figure 1 - A standard 1kg mass kept from changing in the atmosphere by being encased in a vacuum

	Name		deca-	hecto-	kilo-	mega-	giga-	tera-	peta-	exa-	zetta-	yotta-
Multiples	Symbol		da	h	k	M	G	T	P	E	Z	Y
	Factor	1	10^1	10^2	1000	1000000	1000000000	10^{12}	10^{15}	10^{18}	10^{21}	10^{24}
	Name		deci-	centi-	milli-	micro-	nano-	pico-	femto-	atto-	zepto-	yocto-
Fractions	Symbol		d	c	m	µ	n	p	f	a	z	y
	Factor	1	10^{-1}	0.01	0.001	0.000001	0.000000001	10^{-12}	10^{-15}	10^{-18}	10^{-21}	10^{-24}

Figure 2 - Standard Prefixes

At iGCSE level you will only need to understand those that are not shaded out and how to covert to and from the base units using these prefixes. The table highlights that prefixes are either used to describe larger units (for example 1000 metres can be called 1kilometre – 1 kilometre is 1000 metres and so is larger than 1 metre). Larger units are called "Multiples".

RULE 1: To convert TO a Fraction, you divide by the factor. To convert FROM a Fraction, you multiply by the factor.

E.g. 1100m is the same as 1100/1000 which is 1.1 kilometres (converting to a Multiple requires dividing by the factor)

E.g. 3.6 kilometres is the same as 3.6 x 1000 metres, or 3600 metres.

Some prefixes are used to describe smaller physical quantities. An atom, for instance is approximately 0.0000000001 metres across. This would take a long time to write too often and I might make a mistake in the number of 0s I have added.

RULE 2: To convert TO a Fraction, you divide by the factor. To convert FROM a Fraction you multiply by the factor.

Using the Fractions part of the table above it is possible to express this in nano-metres instead. To get to this I will divide by its factor or 0.000000001:

$$\frac{0.0000000001}{0.000000001} = 0.1 nm$$

It is a lot easier to write 0.0000000001m as 0.1nm instead.

Doing the opposite, I can covert 3cm to metres by multiplying it by the factor:

$$3cm * 0.01 = 0.03m$$

I know now that 3cm is the same as 0.03 metres in length.

The overall rule then is:

*Converting **TO** a Prefix from a base unit uses a division, converting **FROM** a prefix to a base unit uses a multiplication*

1.2 Scientific notation

Physics is full of numbers, such as the speed of light and Planck's constant. These are numbers that are either big (speed of light is 300000000 metres each second) or small (Planck's constant is 0.0000000000000000000000000000000006626068). Imagine having to write number this big or this small too often.

Luckily, mathematicians have devised a simpler way of doing this by realising that each time there is a 0 at the end of a large number, it is the same as multiplying with 10, and if a small number is prefixed with a 0, it is the same as dividing by 10.

So 100 is the same as 10 x 10, or 10000 is the same as 10 x 10 x 10 x 10. From mathematics you will be familiar with powers and will quickly notice that 10x10 is 10^2 and 10 x 10 x 10 x 10 is 10^4. We can use this to more easily express large numbers like this:

Eg. 548000 ⟶ 5.48×10^5 (the power counts the number of times 10 is multiplied)

0.0049 ⟶ 4.9×10^{-3} (the power counts the number of times 10 is divided)

500 ⟶ 5 $\times 10^2$

0.0003 ⟶ 3 $\times 10^{-4}$

*Large numbers are expressed as **10** $^{a\ positive\ number}$ and small numbers are expressed as **10** $^{a\ negative\ number}$.*

Using this system, also known as "Standard form", the speed of light is 3×10^8 metres each second and Planck's constant is 6.626068×10^{-34}. Much easier!

1.3 Measurement

Measurement of Mass

Mass is the amount of "stuff" in an object. It might be thought of how big and how heavy the smaller particles are in an item, and how many of them there are. The bigger, the heavier and the more particles there are in an object, the more massive it is.

Mass is measured using either a **digital mass balance** or a **beam balance**

Figure 3 - Mass Balance

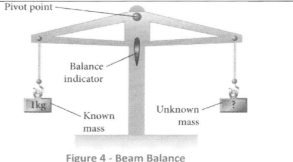

Figure 4 - Beam Balance

MASS BALANCE	BEAM BALANCE
The digital mass balance shown above shows how much mass there on the pad. Before placing an object on it, it is important to "zero" is. There is a button on most of these devices to do that	The balance shown above uses known masses on one side of a pivot to balance against an unknown mass. This can take some time to get used to using, but can be very accurate. **This kind of apparatus can be used to measure masses on different planets**, whereas the mass balance will not be accurate beyond the Earth's surface.

Mass is slightly different.

Nearly all base units are expressed in the form without a prefix, but the **kilo**gram has the prefix in it already; "kilo". Note that the kilogram IS the base unit, not the gram. This makes converting a little more awkward as you have to consider this carefully when converting. The initial definition of a kilogram was the mass of 1litre of water, of 1000cm³ of water, however this soon became rather less than useful as water changes density (how much mass in a certain volume) at different temperatures.

Look at the table below and how it uses the information in Figure 2 - Standard Prefixes.

Mass		Whole number	Standard Form
1 Kilogram		1 kg	
1 Gram	1 g	1/1000 kg	10^{-3} kg
1 Milligram	**1/1000 g**	**1/1000000 kg**	10^{-6} kg

Measurement of Time

Time measurement was originally devised as a way of splitting a day, months and years into smaller, easy to divide parts. The year is the time it takes the Earth to orbit the sun and so it is important to know when it starts and stops for harvesting and sowing. The month was originally the time it took for the moon to orbit the Earth, and so understanding this helped understand tides for fishing. The day is the time it takes for the Earth to spin around its axis once and understanding what time of day it is helps understand when to start working hunting, and feeding the cattle.

In the early days it made sense to break the day into quarters, 6ths and so on, and so the day was broken into two lots of 12 hours, or 24 hours. 12 from midday to midnight and 12 from midnight to midday, and each hour was broken down into minutes, 60 in each hour. Even a minute is quite a long time, and so these were broken down into seconds.

1 minute is 60 seconds

1 hour is 60 minutes, each minute has 60 seconds, so 1 hour has 60 x 60s, or 3600s

1 day has 24 hours, or 24 x 60 minutes, 1440 minutes, or 1440 minutes of 60 s each, or 86400s.

Table 2 - Longer time periods expressed in seconds

1 second	1 minute	1 hour	1 day
1 second	60 seconds	3600 seconds	86400 seconds

This way of working out minutes and hours in seconds is essential in physics.

Minutes, hours and days are longer periods of time than the base unit, second. It is also possible to measure shorter time periods than seconds. These follow the method in Figure 2 - Standard Prefixes shown above. So, 1 millisecond is 0.001s.

To measure time you can use clocks, watches or timers, such as stop watches. These are either digital (with numbers) or analogue (with dials that move smoothly)

Figure 5 - Digital stopwatch	Figure 6 - Analogue stopwatch
Digital stopwatch. This watch shows the time passed since starting it as 12 minutes, 0 seconds and 270 milliseconds	Analogue stopwatch. This watch shows the time passed since started as 17 minutes (the lower dial) and 9 seconds (the outer dial).

As mentioned above, shorter periods than the base unit uses the standard factoring method in Figure 2 - Standard Prefixes above and some examples are shown below.

Table 3 - Short periods of time expressed in seconds

Time	Fraction	Scientific notation
1 millisecond (ms)	1/1000 s	1×10^{-3} s
1 microsecond (µs)	1/1000 000 s	1×10^{-6} s
1 nanosecond (ns)	1/1000000 000 s	1×10^{-9} s

Measurement of Length

Length is defined nowadays using how far light passes in a particular fraction of a second (about 3×10^{-8}s), but changed over time. At one point it was based on the length of a pendulum and then as $1 \times 10^{-7\text{th}}$ of the distance following the curvature of the Earth from the equator to the North Pole. Both these methods were very impractical and so a simpler method was devised that used a standard length of a very inert (unchanging) substance that each country bought and was used to define 1 metre. This was called the *mètre des Archives* and is made out of a Platinum-Iridium mixture.

Figure 7 - A picture of the US standard metre bars

Methods of measurement

In physics, length can be as small as being inside an atom, or as big as the Universe itself. Measuring very small sizes and very large sizes are performed using very specialist methods that involve complex instruments, such as telescopes or microscopes. At GCSE, however, you will need to know about how to measure more everyday items. These are measured using a **ruler**, **calliper** or **micrometre**.

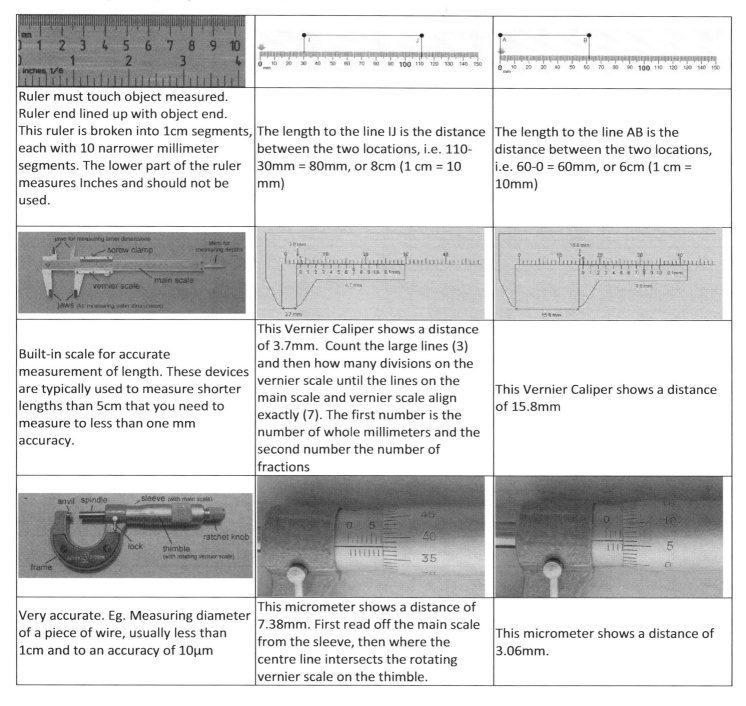

Ruler	Caliper example 1	Caliper example 2
Ruler must touch object measured. Ruler end lined up with object end. This ruler is broken into 1cm segments, each with 10 narrower millimeter segments. The lower part of the ruler measures Inches and should not be used.	The length to the line IJ is the distance between the two locations, i.e. 110-30mm = 80mm, or 8cm (1 cm = 10 mm)	The length to the line AB is the distance between the two locations, i.e. 60-0 = 60mm, or 6cm (1 cm = 10mm)
Built-in scale for accurate measurement of length. These devices are typically used to measure shorter lengths than 5cm that you need to measure to less than one mm accuracy.	This Vernier Caliper shows a distance of 3.7mm. Count the large lines (3) and then how many divisions on the vernier scale until the lines on the main scale and vernier scale align exactly (7). The first number is the number of whole millimeters and the second number the number of fractions	This Vernier Caliper shows a distance of 15.8mm
Very accurate. Eg. Measuring diameter of a piece of wire, usually less than 1cm and to an accuracy of 10μm	This micrometer shows a distance of 7.38mm. First read off the main scale from the sleeve, then where the centre line intersects the rotating vernier scale on the thimble.	This micrometer shows a distance of 3.06mm.

Distance	Comparison	Standard notation
1 Kilometre (km)	1000 m	10^3 m
1 Metre (m)	1 m	
1 Centimetre (cm)	1/100 m	10^{-2} m
1 Millimetre (mm)	1/1000 m	10^{-3} m
1 Micrometre (µm)	1/1 000 000 m	10^{-6} m

Parallax errors

When reading off any measurement you must always try to read the instrument from a directly in front of it, at a right angle to the division, or else you might end up with an error, as shown in the diagram. This type of error is called a parallax error:

Parallax: The difference between the perceived and the actual position of an object

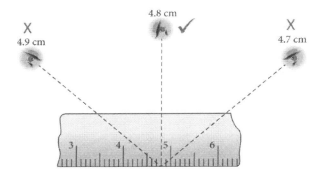

Figure 8 - Parallax

1.4 Questions:

1. Write down the value of:
(a) 1564 mm in m
(b) 1750 g in kg
(c) 3.65×10^4 g in kg
(d) 62 µs in s
(e) 6.16×10^{-7} mm in m

2. The pages of a 500 page book have a mass of 2.50 kg. What is the mass of each page:
(a) In kg   (b) In mg.

3. Arrange the below units in 3 columns in the table below:
km   µg   t   nm   kg   µs   mg   ns   g   mm

MASS	LENGTH	TIME

4. In each of the following pairs, which quantity is the larger?
(a) 2 km or 2500 m
(b) 2 m or 1500 mm
(c) 3000 kg or 30,000,000 µg

(d) 40 µs or 4 ns

5. Which of the following is/are correct?
(a) one mg = one million g
(b) 1000 mg = 1g
(c) 1000,000 µg = 1g
(d) 1000,000 mg = 1 kg

6. A runner runs a marathon of 36000m in 10800 s. Convert to km/h.

7. The distance between two points is 47 km. Convert to
(a) cm   (b) m   (c) mm   (d) µm

8. The speed of light is 3×10^8 m/s. It travels 15000 km. Calculate the time taken (use the equation speed = distance ÷ time).

9. Calculate the:
(a) Amount of seconds in a 28 day month.
(b) Amount of kg in a 106000 g object.
(c) Amount of cm in a 3 km length.

Chapter 2. Forces and their effects

2.1 Forces

Nothing happens unless something causes it to happen, this is called a force.

Forces are pushes, pulls or twists – they can cause changes. It is possible to measure these using a **Newton Metre** and forces are measured in **Newtons** (N). One specific type of force is caused by gravity acting on mass, such as the Earth pulling on you, and this is called **Weight**. On Earth the force of gravity is 9.8N/kg. All forces, including weight, cause objects to speed up. This is called **acceleration**.

- Forces push, pull or twist
- Can be measured with a newton-meter
- The unit for force is **Newton** (N)
- **Weight** is a force caused by gravity
- Gravity on Earth is **9.8N/Kg**

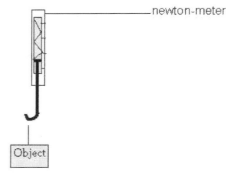

Figure 9 - Newton meter

Figure 10 - Free body diagram

Types of force
- Gravity causes a force on object called **weight**.
- **Thrust**, **Push** and **Pulls** cause acceleration
- **Friction**, such as air resistance and drag cause things to slow down (deceleration or retardation)
- **Lift** is an upward force, opposing weight
- **Tension** is a force that stretches objects
- **Compression** is a force that squashed objects
- **Torsion** is a twisting force
- **Reaction** forces act in the opposite direction to another force

What can forces do?
- The can cause things to speed up, or slow down – **acceleration**
- **Push** or **pull** things
- Change the **size** of an object
- Change the **shape** of an object
- Change the **direction** of the movement of an object

2.2 Weight as a force

Stretching a spring

When a stretching force is applied to a spring it will extend. If twice the force is applied, the extension will also double in length. As the force increases, so does the extension by the same amount. Looking at the extension as a result of being stretched by a force leads us to conclude that the extension is proportional to the stretching force

$$Extension \propto stretching\ force$$

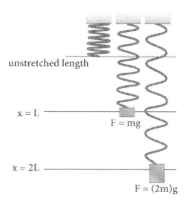

Figure 11 - Hooke's Law - extension is proportional to the force applied

This is called **Hooke's Law**.

Robert Hooke (1635-1703) was a genuinely eccentric genius. He has mainly been credited for his work on the stretching forces, but was also responsible for microscopes, astronomy, naming "cells" in organisms and for research on gravity. He was disliked by many and, unlike most other famous scientists of his era, no one knows what he actually looked like and his portrait in the royal society was, conveniently, lost. All paintings of him are just based on descriptions in writings, or wrong.

As the graph to the below shows, the extension and the force, when plotted, create a straight line. From mathematics you know that this is called **"directly proportional"** and can be expressed as y=kx+m

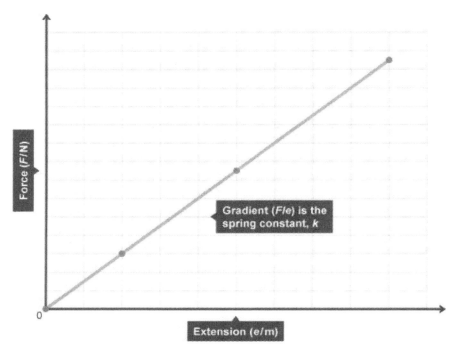

Figure 12 - Graph of a string while it is being stretched with its Elastic Limit

However, if you pull a spring too far it will not return to its original shape. The point at which it has been stretched too far is called the **Elastic Limit**. While the spring can return to its original shape, it obeys Hooke's "directly proportional" law, but once it has stretched beyond this, it no longer obeys it, as seen in the next graph.

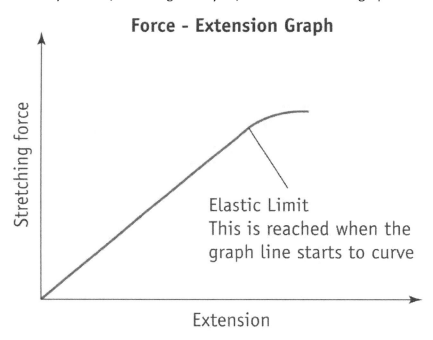

Figure 13 - Hooke's law is observed UNTIL the Elastic Limit and then the line is no longer straight.

As Figure 13 - Hooke's law is observed UNTIL the Elastic Limit and then the line is no longer straight. This point is known as the **limit of proportionality**. Beyond this point the spring is deformed (permanently stretched).

Force constant

The force constant, *k*, of a spring described the force needed to cause the same amount of extension and is measured in N/m (Newtons per metre). The force constant can be determined from the graph by calculating the gradient **while within the limit of proportionality**

$$F = k * e$$	Quantity	Name	Units
	F	Force	Newtons (N)
	e	Extension	m
	k	Spring constant	N/m

Measuring the force constant of a spring

All springs behave slightly differently to being stretched. This is because they might be of different materials or just have different thicknesses. Two springs are therefore likely to have different spring constants.

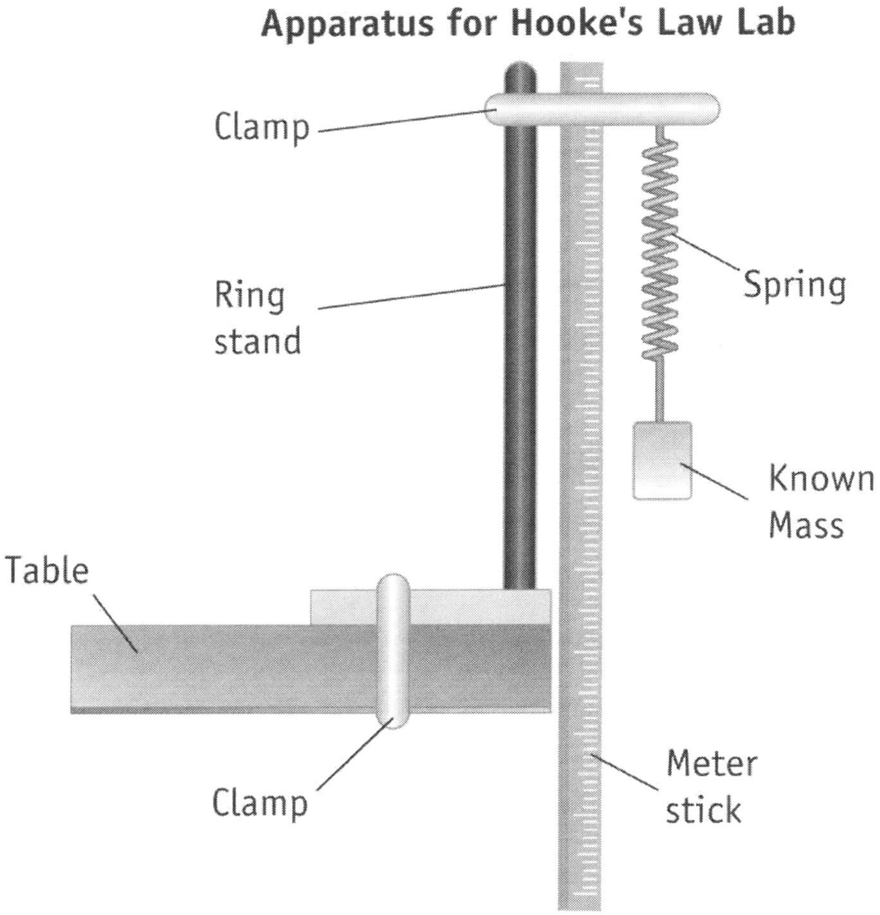

Apparatus for Hooke's Law Lab

Figure 14 - Experimental setup to explore Hooke's Law

Using the apparatus above, you follow the instructions below to draw a graph of extension on the x-axis to the weight (mass) suspended from the spring.

1) Measure the length of the spring on the meter stick
2) Add a weight on spring
3) Record the number of weights in a table
4) Measure the new length of the spring on the meter stick.
5) Record the length in a table next to the weight in step 3
6) Repeat steps 2 through 5 until the spring breaks
7) Calculate a third column by subtracting the length in step one from each measurement of the spring. This is the extension.
8) Draw a graph of extension against weight

Weight (N)	Length on meter rule (cm)	Extension (Length – 12cm from first measurement of spring)
0	12 (for example)	0
1	16	4
2	20	8
3	24	12

Once the graph is completed, this should look like **Error! Reference source not found.**

Worked Example:
A spring is stretched 0.01m by a weight of 2.0N. Calculate

(a) the force constant **k**

(b) the weight **W** of an object causing an extension of 0.08m.

ANSWER:

k = F/e = 2.0N/0.01m = 200N/m

W = Stretching Force F = k x e

= 200N/m x 0.08 = 16N.

2.3 Questions

Calculate k for the following stretches:

a) 0.05m by a force 3N.
b) 1.7m by a force 12N.
c) 20mm by a force 4N.

Calculate W for the following:

a) 150N/m, 1.4m extension.
b) 25N/m, 2.0m extension.
c) 10N/m, 4m extension.

Calculate e for the following:

a) 240N/m, 7N.
b) 1.42N/m, 12.4N.
c) 22500N/m, 68000N.

Chapter 3. Mass and weight

3.1 Differences between mass and weight

Sir Isaac Newton was said to be sitting underneath a tree when he observed an apple fall to the ground. Why did it do this? Why not upwards, or to the side, or, even, just stay where it was? His conclusion was that something was pulling on it. This was called "**weight**" and is not the same thing are "**mass**".

Mass is the stuff we, and all objects, are made out of and is measured in kg (kilograms). As seen above, you measure mass with scales. Mass is a **scalar** – it has only got size, but no direction.

Weight is measured in N (Newtons) and is a force. Weight is measured with a Newton-meter and has both size and direction, which makes it a **vector**.

- A scalar has only got magnitude (or size), but no direction. Examples are mass, speed, density (see later) and temperature.
- A vector has two parts, a magnitude (as in a scalar) and also a direction. Examples of vectors are Force, Velocity (not speed), momentum and acceleration

Figure 15 - Isaac Newton 1642-1727. Discoverer of the theory of gravity, inventor of algebra, the telescope and deductor of the three laws of motion.

Mass	Weight
Measured in kg	Measured in N
Measured with a mass balance	Measured with a Newton-meter
Scalar	Vector
The amount of matter	Mass acted upon by gravity
Not a force	A Force that acts straight down towards the centre of a planet, or other large object.

Different planets have different masses, and so a stronger gravitational pull on an object. Our Moon, for example has a much weaker pull than the Earth on an object. To lose weight, you could therefore just go to the Moon and you will weigh less. You will, however, not have less mass. If you had 80kg of mass on Earth, you still have this on the Moon.

There is simple way of working out your weight on a planet using this equation:

$W = m * g$	Quantity	Name	Units
	W	Weight	Newtons (N)
	m	Mass	kg
	g	Gravitational field strength	N/kg

In this equation, the gravitational field strength changes between planets and stars. On Earth it is on average 9.8N/kg (in most GCSE syllabuses in the UK this is rounded up to 10N/kg). When you use this equation on earth you will use g=9.81 to work out the weight on an object. The table below has examples of g on other planets (you do not need to learn these, but you might have to know how to use them):

Table 4 - gravitational field strengths for the different planets in the solar system

Planet	g - Gravitation field strength (N/kg)
Mercury	4
Venus	9
Mars	4
Jupiter	26
Saturn	11
Uranus	11
Neptune	14
The Moon	1.6

So, if you wanted to lose weight you could transfer to any over the rocky planets near the Sun. However, if you wanted to weigh more, you could try Jupiter, where you would weigh 2.5 times as much as on Earth.

Worked example

What is the weight of a car with a mass of 1100kg on Earth? What is its weight on the moon?

For Earth, g is 9.8N/Kg, and so its weight is:

W=1100 * 9.8 = **10780N**

For the Moon, g is 1.6, so its weight there is:

W=1100 * 1.6 = **1760N**

The gravitational field strength
Little g is different, depending on where you are. On Earth it has an average value of about 9.81N/Kg, but this also varies throughout the Earth. In general, the closer to the centre of a body (like a planet, or a star) you are the stronger the pull of gravity is, the higher your weight is. So, on Earth, where the poles are squashed

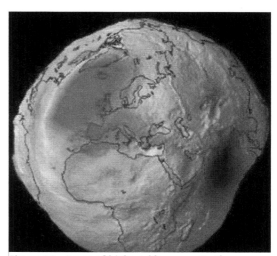

Figure 16 - Areas of high and low g on Earth

towards the centre of the earth, you weigh more than on the equator. At the poles g is 9.83, whereas on the equator it is closer to 9.78. If you want to lose weight, just move to the equator.

However, it is more interesting than that as it also depends on where the densest (see later) elements in the Earth's core are concentrated. Figure 16 - Areas of high and low g on Earth, analysed by the GOCE satellite, shows (in a highly exaggerated way) where the gravitations field strength is strongest (the red areas).

The gravitational field strength is the acceleration experienced around a body, and so is also expressed as m/s^2. If an object falls on Earth, it will accelerate with the rate of the gravitational field strength. So, on Earth, if a pen was dropped from a height, it will accelerate with a rate of $9.81 m/s^2$.

3.2 Vectors and Scalars

Scalar	Vector
has SIZE only	has SIZE and DIRECTION
Speed, volume, mass, density	Velocity, displacement, acceleration, force
Adding two scalars simple involves adding the two sizes.	Adding vectors requires understanding the direction of the two quantities.

Adding vectors and scalars

Vectors have both a size and a direction. In mathematics and in physics it is easy to represent these using arrows. The direction the arrow points describes the direction of the vector and the length of the arrow describes the size of the vector. In the example below, the arrows represent two forces of the same size acting on an object.

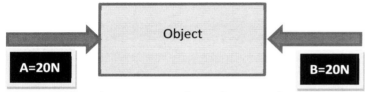

This could be a box, or a car, that is being pushed at both ends by two persons, equally strong. A and be oppose each other. Choosing right as the positive direction, A+B becomes 20N + -20N, which is 0. These forces are **balanced**.

The next diagram shows how the same body is acted upon by two different forces, opposing each other. As a result of the force from the left being the larger one, the object will start moving towards the right. This is because there is a net force to the right. When there is a net force, it is said to be **unbalanced**

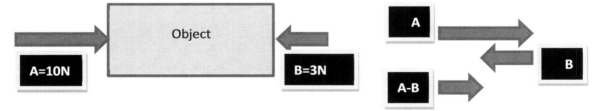

Here we choose right as positive, so A=10N, B=-3N, A+B is 10 + -3 = 7N. Since the result is positive, it is to the right.

The net force can the determined by adding force A to force B (remembering to choose your coordinate system), as done to the right of the diagram. It is also possible that two forces assist each other, as in two people pushing a car. When the net force is zero, it is said that the forces are **balanced.**

Having two people push the car provides a higher force and so you add the forces as is done to the right of the diagram. The net force is A+B, 10N + 3N = 13N.

> *Forces that oppose each other may produce a net force, that is the difference between the two, and its direction is the same as the larger force.*
> *Forces that act in the same direction, and that are parallel add to each other, becoming stronger. Its direction is the same as the forces and its size is the sum of the forces.*
> *Another way of saying net force is **"resultant force"***

Adding Vectors Which Are Not In Line

Most forces do not come straight at each other; instead they might be at strange angles. Most GCSE syllabuses do not explore this part of the force addition. Check yours before learning this.

Two forces at right-angles

If the two forces work at right angles to each other they form a right angled triangle and so we can use Pythagoras theorem to work out the resultant force.

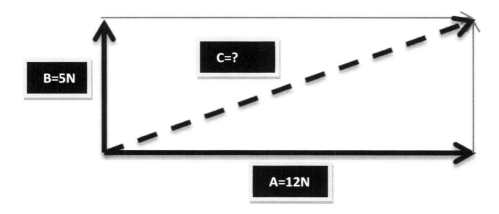

Using Pythagoras:

$$A^2 + B^2 = C^2$$

Resolving for C, we get:

$$C = \sqrt{A^2 + B^2} = \sqrt{25 + 144} = \sqrt{169} = 13N$$

Two forces at a non-right angle

When two angles are not at a convenient right angle

1) Use graph paper
2) Use a protractor
3) Choose a suitable scale, e.g. 1cm = 10N.
4) Draw the two forces, to scale, connecting at a common vertice

E.g. two forces act at 30 degree to each other, one is 10N, the other is 5N. Work out the resultant force.

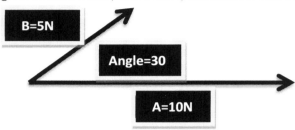

5) Create a parallelogram as follows

6) Draw a line from the vertices between A and B, and that form opposite – this is the resultant force

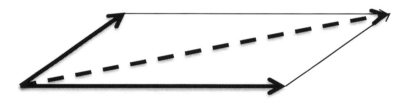

7) Measure the dotted line and use the scale you chose to calculate the resultant force.

3.3 Questions:

1. Draw 2 forces (100N and 140N) at right-angles on graph paper (use a scale of 1cm = 20N). Calculate the resultant force.
2. Repeat the above for 60N and 180N.
3. Draw the below accurately on graph paper. Use a protractor. Calculate the resultant. (Use 1cm = 10N).

3.4 Forces can change the motion of an object

Isaac Newton penned three laws that describe how forces affect each other and the motion (movement) of an object. You need to know the first two well. The last one is less relevant at GCSE level, but is explored in detail at A-level.

3.5 Newton's First Law

> *First Law*: *"If an object is at rest it will stay at rest unless an external force acts on it."* (inertia). *"If the forces on an object are BALANCED it will move in a straight line at CONSTANT VELOCITY."*

Examples
Let us first explore a force that is horizontal – going across.

Unbalanced force:	**Balanced forces:**
This is when forces to not cancel each other. There will be a net force, and then the object will change its speed in the direction of the net force.	This is when two forces that oppose each other leave a net force of 0.
Weak force is to the right. Strong force is to the left. The object. Net force is to the left.	Some force to the right. Same amount of force to the left. The object. Net force is zero.
The push from the left is smaller than the push from the right. There will be a net force equal to the difference between the two forces and the object will start to accelerate to the left (in the direction of the net force)	The pushes from left and right are exactly the same in size and direction. There will be no net force. As a result the object will keep the same velocity. This can mean that it stays still or stays at the same speed and direction
An example: A car where the thrust from the engine is higher than the resisting friction forces from the road and the air	An example: A car driving straight ahead where the thrust from the engine equals that of the air resistance and friction from the road

Next we explore that happens when the forces are vertical where gravity is acting on an object

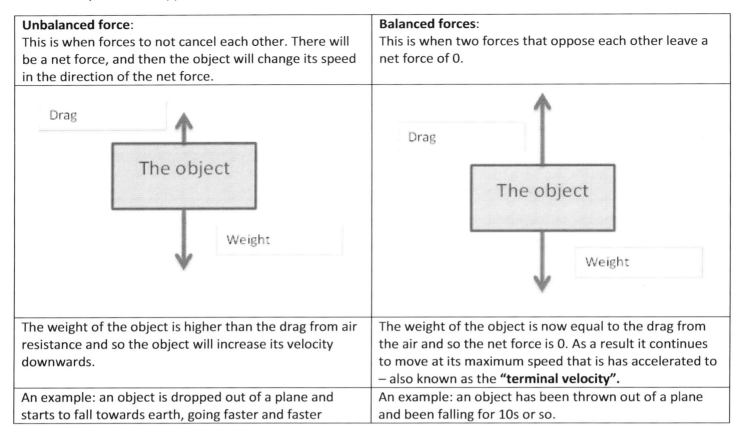

Unbalanced force:	Balanced forces:
This is when forces to not cancel each other. There will be a net force, and then the object will change its speed in the direction of the net force.	This is when two forces that oppose each other leave a net force of 0.
The weight of the object is higher than the drag from air resistance and so the object will increase its velocity downwards.	The weight of the object is now equal to the drag from the air and so the net force is 0. As a result it continues to move at its maximum speed that is has accelerated to – also known as the **"terminal velocity"**.
An example: an object is dropped out of a plane and starts to fall towards earth, going faster and faster	An example: an object has been thrown out of a plane and been falling for 10s or so.

The two sets of forces are different, but they interact with each other in the same way. To find out if a force is acting on an object in Newton's first law we have to consider that there may be several forces and we need to determine what size and direction the net force is acting.

3.6 Newton's Second Law

Second Law: "Unequal forces produce acceleration according to a=F/m"

As seen above in the unequal horizontal and vertical scenarios, when unequal forces are in action, there will be some form of acceleration as a result.

The acceleration will double if the force, F, doubles and so the acceleration, a, is directly proportional to the net force:

$$a \propto F$$

However, the more massive an object is, the more difficult it will be to produce acceleration. This means that the acceleration is inversely proportional to the mass:

$$a \propto 1/m$$

Putting these two together we find that Newton's second law can be expressed mathematically as:

$F = m * a$	Quantity	Name	Units
	F	Force	Newtons (N)
	m	Mass	kg
	a	Acceleration	m/s²

Using a visual method of showing this, we can use the diagrams above to show that there is acceleration occurring by showing that there is a net force acting:

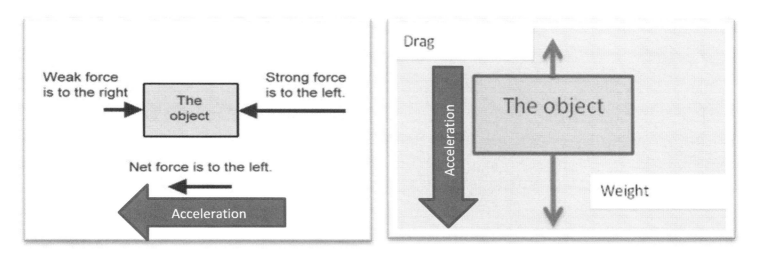

Using our knowledge of forces from above, we can look at the situation where a bus is acted upon by two forces; A) the thrust from the engine and B) air resistance

In this case we can deduce:

1) It is an unbalanced force since there is a resultant force
2) Resultant force above = 200 – 70 = **130N**.
3) The resultant force will be towards the left
4) The bus will therefore accelerate (increase in velocity) in that direction

From above we also learned that the greater the mass, the more force is required to accelerate the bus, in other words, mass resists acceleration. Newton's second law allows us to work out how much an object accelerates based on its mass and the net force acting on it.

Worked Example 1

Calculate the force needed to accelerate a mass of 12kg at 5m/s².

$$F = 12 \times 5 = 60N.$$

Worked Example 2

Calculate the acceleration of a mass of 400kg if a force of 1600N acts on it.

$$a = F \div m = 1600 \div 400 = 4 \text{ m/s}^2.$$

3.7 Questions

(a) F = 5000N   m = 250kg.   a = ?

(b) a = 22m/s²   m = 6kg.   F = ?

(c) F = 12000N   a = 52m/s².   m = ?

3.8 Newton's Third Law

Third Law: "For every force (or action) between two bodies there is always an equal but oppositely directed force (or reaction) (of the same type)"

Newton's Third law is the most misunderstood part of the laws of motion. It is best to understand it by thinking of there always being **pairs** of forces when one object acts on another object. Look at the table below.

| 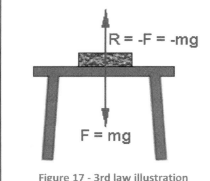 Figure 17 - 3rd law illustration | The book on the table exerts a force, due to gravity (W=mg) on the table. Since the book is not accelerating, i.e. it is at a constant velocity (in this case that velocity is 0 – it is not moving), there is no NET force. That means that there has to be a force acting on the book from somewhere else. In this case this force is from the table on the book. We have therefore identified a pair of forces:
 1) A force on the table from the book (mg)
 2) A force on the book from the table (-mg)
 This is the simplest form of the third law in action |

An easy experiment can be performed in a school that shows this in action, using two persons on skateboards, or roller skates. Person 1 pushes against person 2. Unless the third law was true, only person 2 should move away. However, as a result of person 1 exerting a force on person 2, person 2 exerts exactly the same force back and so both persons move backwards from the other, as seen to the right. The force pairs seen (F_A and F_B) are equal. The acceleration is calculated using a=F/m, and so if person A is heavier than person B, they will have a lower acceleration (since F= F_A and F=F_B)

Figure 18 - Force pairs in skaters

3.9 Questions

1) Calculate the acceleration of an object if it increases its velocity from 71m/s to 162m/s in 15s.
2) Calculate the force if an object has a mass of 56kg and an acceleration of 245m/s^2?
3) Calculate the mass of a body accelerating from 45m/s to 265 m/s in 21s with a force of 23,452N.
4) What force is needed to accelerate a mass of 12kg at 5m/s^2? The same force acts on another mass and it accelerates at 6m/s^2. What is its mass?
5) Calculate the force acting on a mass of 45kg accelerating from 20m/s to 500m/s in 12s.
6) A body of mass 1.2g accelerates from 26m/s to 79m/s in 34s. Calculate the force.
7) A body of mass 265kg decelerates from 30m/s to 2m/s in 7s. Calculate (a) the deceleration and (b) the force required.
8) A car of mass 2/100,000g is acted on by a force of 24,000N. Calculate the acceleration.
9) A force of 240N acts on a mass of 240g. Calculate the acceleration.
10) A missile decelerated from 8000m/s to rest in 2s. Calculate its mass of the force on it is 120,000N.

Chapter 4. Speed, velocity and acceleration

It is important in all lines of science to understand the difference between a scalar and vector. We discussed how a force is a vector and how mass is a scalar. In this chapter we will explore two concepts that are very closely related, but which are different in that one is a scalar (speed) and the other a vector (velocity).

Both are measured in metres per second (m/s), but are subtly different. Acceleration is, as we will see, based on the velocity of an object, not its speed.

Speed	Velocity
A Scalar (has only got a size)	A Vector (has got both size and direction)
$$Speed = \frac{distance}{time}$$ A car travels 20m in 5 seconds. Find (a) its speed (b) how long it takes to travel 75m. Speed = d/t = 20/5 = 4m/s. t = d/speed = 75/4 = 18.75s.	$$velocity = \frac{displacement}{time}$$ $$v = s/t$$
Measured in metres/second (m/s)	Measured in metres/second (m/s)
Speed does not change if the object travelling changes direction	The velocity changes is an object changes direction

As we will see below, exploring motion over time, you might know the speed at a certain point in time (by, for instance, checking your speedometer in the car), but your **average speed** between two points might be very different. The average speed is calculated using

$$Average\ Speed = \frac{total\ distance\ travelled}{total\ time\ taken\ to\ travel}$$	

As such, the average speed is reduced when a car is forced to stop for a while at traffic lights, accelerates to overtake and travels at a constant speed along a motorway.

$$Speed = \frac{distance}{time}$$	The scalar version of the motion equation
$$velocity = \frac{displacement}{time}$$	The vector version of the motion equation

4.1 Acceleration

The acceleration of an object is how rapidly the VELOCITY of an object changes. Typically this involves changing the size of the velocity, e.g. from 10m/s to 20m/s, but since it is based on the velocity, and because a velocity is a vector, acceleration also occurs when the direction of motion changes. This occurs, for instance, when a car goes around a corner.

$$Acceleration = \frac{change\ in\ velocity}{time\ taken\ to\ change\ the\ velocity}$$

More easily written as:

$$a = \frac{v - u}{t}$$

Quantity	Name	Units
a	Acceleration	m/s²
v	Final velocity	m/s
u	Initial velocity	m/s
t	Time taken to change the velocity between u and v	s

Acceleration is measured in m/s, per second, or m/s/s, more easily written as m/s². It is easy to think of this in terms of the change in m/s each second. So, if the acceleration is 5m/s² you could read this as a "change in velocity of 5m/s, each second".

Remember that velocity is a vector, and so acceleration is a vector also, and so has size and direction. When using Newton's second law, F=m*a, both F and a are vectors.

Example
Calculate the acceleration of an object whose velocity changes from 2m/s to 6m/s in 5.6s.

$$a = \frac{v - u}{t} = \frac{[6 - 2]}{5.6} = 0.71 m/s^2$$

4.2 Questions

1) Calculate a for a change of 8100m/s to 40m/s in 22s.
2) Calculate a for a change of 572m/s to 59000m/s in 11s.
3) Calculate a for a change of 0.14m/s to 7.9m/s in 2m.
4) Calculate a for a change of 22m/s to 4.9m/s in 8m 24s.

4.3 Using graphs to describe motion

Distance-time graphs (plotting how far something has moved against the time taken)

In these graphs, the x-axis is the time taken and the y-axis describes how far something has moved. If there is a change in the distance, the object has changed location and has therefore moved. In order to move it has to have a speed between those two positions.

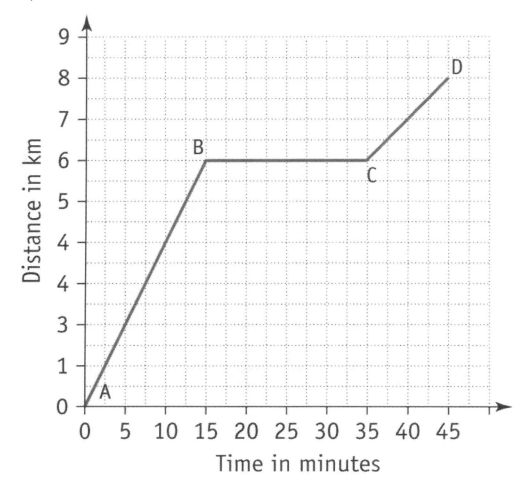

Figure 19 - Distance time graph

In Figure 19 - Distance time graph above, an object starts are a position A and changes to position B, 6km further away. By reading off the Y-axis, you will note that this journey took 15 minutes. We can calculate the speed between these two location by, firstly converting to base units – so 6km is 6000m, and 15 minutes is 60x15s = 900s.

$$speed = \frac{distance}{time} = \frac{6000m}{900s} = 6.7 m/s$$

Between points B and C, there is no change in distance and so the object was not moving, and so the speed is 0. At point C it starts moving again until point D. Here we can again calculate the speed.

$$speed = \frac{distance}{time} = \frac{8000 - 6000m}{(45 - 35) * 60} = \frac{2000m}{600s} = 3.3 m/s$$

It is easy to estimate a speed from the graph by looking at the gradient of the line.

1) The gradient provides the speed
2) When the gradient is 0, the speed is 0
3) The steeper the gradient, the higher the speed

The **average speed** travelled over the 45 minutes is different than either of these two values

$$average\ speed = \frac{total\ distance\ travelled}{total\ time\ taken} = \frac{8000m}{45*60} = \frac{8000m}{2700s} = 3.0 m/s$$

Figure 19 - Distance time graph is a very simple diagram which might show a car travelling at 6.7m/s, then stopping at a shop for 20 minutes, then travelling at 3.3m/s. It does not tell us about changing speed or the direction it is travelling.

A more accurately drawn graph would show when the car accelerates or decelerates at points A, B, C and D. The way the graph is drawn assumes that the change of speed is instantaneous, and that is impossible.

The following graph shows a simple journey, but with a more accurate way of showing the change in speed.

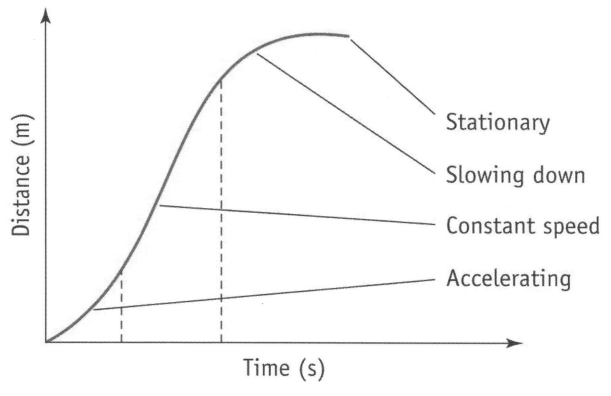

Figure 20 - distance time graph showing changes in speed

Starting at time=0, the car changes speed from stationery to a constant speed. The constant speed is shown by a straight line. Acceleration is shown by a curved line, increasing in its gradient (remember that the gradient represents the speed, so an increasing gradient indicates an increasing speed). Later, the car decelerates until the gradient is 0; the car slows until it stops.

4) An increasing gradient indicates acceleration
5) A decreasing gradient indicates deceleration

Displacement

Displacement is different from distance in that distance measures how far an object travels, but is not interested in how far from the origin it has gone, nor the route taken. Take the picture below, **Error! Reference source not found.**, t he distance between A and B for the car is 25km, but the trains can go directly between the two, using the shortest distance. In this case the cities have a displacement of 10km, but travelling by car it is a distance of 25km.

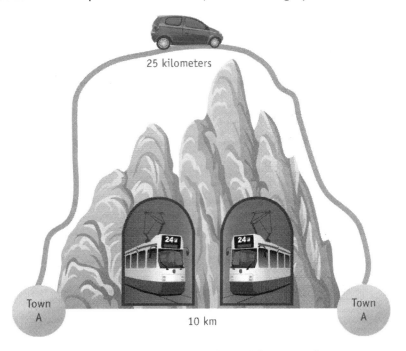

Figure 21 – Displacement=10km and distance=25km

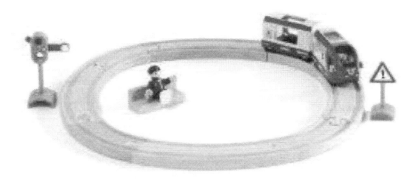

Figure 22 - displacement = 0

Distance is only interested in how far one has to travel to get from A to B. Displacement is a straight line between the destination and the starting point. Most extreme in the case is if a model train, on a circular track, 3m in circumference, started and ended at the same point, the displacement would be 0, but the distance travelled would be 3m.

Velocity is measured using displacement, and, as acceleration is measured using acceleration, acceleration also uses displacement (this part of physics is explored more at A-level, and further down under mechanics, SUVAT).

Velocity-time graphs
When an object changes its velocity over time, we can plot a graph where the x-axis is the time and the y-axis is the velocity. This type of graph is in fact very useful in describing motion.

Velocity - Time graph

Figure 23 - Velocity - time graph – the areas under the graph shows the distance travelled. Use standard area calculations to work out.

Remember that the acceleration can be calculated using

$$a = \frac{v - u}{t}$$

Using this equation we can work out the accelerations for the three regions, A through C. For A, the velocity at the start, u=0, and the end velocity, v=20m/s, the time taken is 10s.

$$a = \frac{v - u}{t} = \frac{(20 - 0)m/s}{10s} = \frac{20m/s}{10s} = 2m/s^2$$

For region B there is no change in velocity, so the object is at a constant velocity. For region C, the starting velocity, u=20m/s, the end velocity, v=0, and the time taken is 40s.

$$a = \frac{v - u}{t} = \frac{(0 - 20)m/s}{40s} = \frac{-20m/s}{40s} = -0.5m/s^2$$

In region A, the acceleration is **positive** ($2m/s^2$) and so the object is increasing its velocity. In region B, the acceleration is **negative** (-$0.5m/s^2$); the object is **decelerating**. This is also called **retardation**.

We can therefore tell from the graph that:

1) The gradient is the acceleration. A negative gradient is deceleration
2) A flat line is constant velocity
3) The steeper the gradient, the higher the acceleration (or deceleration).

What is not so obvious is that the area underneath the graph is the distance travelled.

We can use simple area calculations to explore the distances travelled over A, B and C. A is a triangle, and so the area under it is ½*height*width:

$$\frac{1}{2} * 20 m/s * 10s = 100m$$

B is a rectangle and the area is height * width:

$$20 m/s * 20s = 400m$$

C is a circle and so uses the same as area A:

$$\frac{1}{2} * 20 m/s * 40s = 400m$$

The total distance is therefore 100+400+400=900m.

4) The area under a velocity-time graph is the distance travelled

In some exam questions you are asked to describe motion from a graph. Here is a graph with the terms you should use:

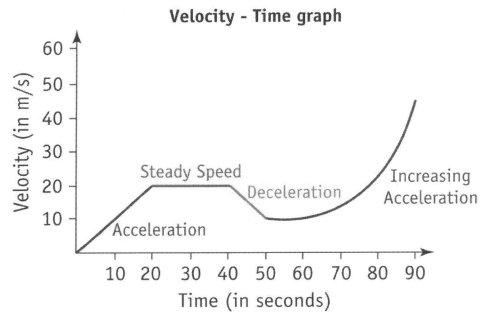

Figure 24 - Terms used to describe a velocity-time graph

5) A curve on graph indicates a change of acceleration. It either increases when the gradient increases, or decreases when the gradient decreases.

4.4 Stopping distances

Imagine seeing a cat run across a road ahead of a car. Now your mind has to process this information and make an informed decision. Those studying biology will know what the sensory nerve from the eyes will have transferred the signal to the CNS (Central Nervous System) where the brain is now processing it. If the brain now decides that the car is about to hit the cat it will inform the motor neurons to control the muscles (the effectors) to move the foot from the accelerator to the brake and press it hard. The time is takes to make this decision and move the foot is called the **reaction time** and the distance moved during this time is the **Thinking Distance**.

Once the brake is pressed, the car will decelerate (negative acceleration). This deceleration takes places as the brake pads squeeze against a disc on the wheel, causing intense friction. The time is takes to stop the car is the **braking time** and the distance moved is the **braking distance**.

Typical Stopping Distances

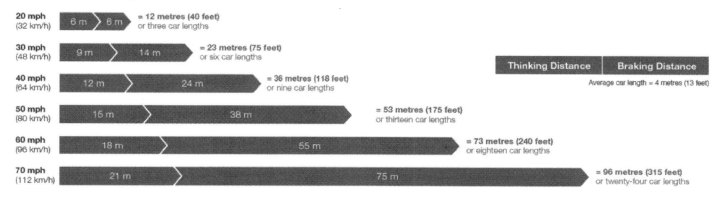

Figure 25 - standard government calculations for stopping distances

The thinking distance is calculated using standard speed equation:

$$s = \frac{d}{t}$$

Using this equation we can work out what the reaction time is that the government assumes we have, using distance = 6m, and the speed is 32000m in 3600s (32km = 32000m, 1hour = 3600s).

$$t = \frac{d}{s} = \frac{6}{\frac{32000}{3600}} = 0.675s$$

The braking distance is calculated using the acceleration equation shown above, and with that we can work out the rate of acceleration that is assumed to take place.

The braking distance can be affected by poor tyres, bad brakes, a slippery road or a poor road surface, whereas the thinking distance is affected by the ability of the driver to respond. Drugs, alcohol, sleepiness and age might affect this.

4.5 Questions

1) Plot on graph paper the below data for a racing car:
   SPEED(m/s):   0 10 20 29 37 50 59 64 65 65
   TIME(s):   0   1   2   3   4   6   8 10 12 14

   What was its acceleration at (a) 13s? (b) 1 second?

2) The velocity v, of a car varies with time, t:
   t(s):   0 5 10 15 20 25 30 35 40   45 50
   v(m/s): 0 5 10 15 15 15 15 11 7.5 3.5 0
   a) Plot a graph of **v** against **t**. Describe the motion between 0-15s, 15-30s, 30-50s.
   b) Calculate the acceleration during 0-15s.
   c) State the forces on the car at constant velocity. What is the resultant?

3) A train takes 1 min to travel between stations. The train accelerated from rest to 25m/s in 20s, travels at this constant speed for 30s before coming to rest after constant deceleration.

   a) Draw a graph to represent the motion.
   b) Calculate: the time the train is traveling more than 20m/s and the retardation of the train.

c) The braking force to stop the train if the mass of the train is 100,000kg.

4.6 Circular Motion

As per Newton's First law of motion, an object will stay at the same velocity unless acted upon by a force. This means that it will neither change its speed nor its direction unless there is a net force acting on it. In order to make an object go in a circle, then, there must be a force acting inwards, causing it to turn.

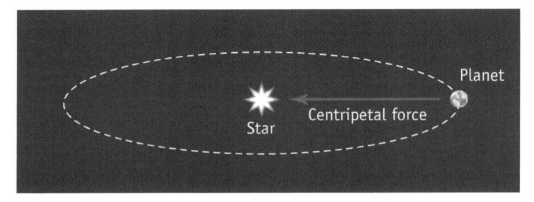

Figure 26 - Centripetal force acting on a planet from the gravity of a star.

This inward force is called **centripetal** force, and examples are:

1) gravity keeping planets orbiting a star
2) a wire on the ball in a hammer
3) the string attached to a ball moving in a circle

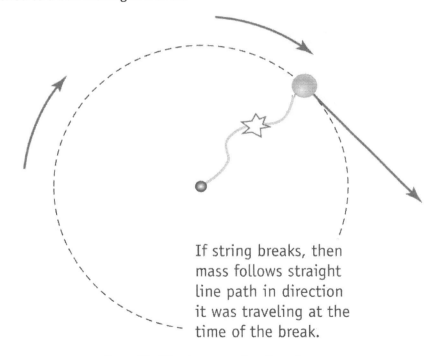

Figure 27 - What happens when the string snaps

Without this force, the object would continue moving in a straight line along its original velocity vector, at the same speed. So, if an object is moving in a circular motion at the end of a string, there is an inward force (experienced as tension in the string). If the string snaps, the object will keep going in a straight line, following the velocity vector at the location in the circular path when the string snapped. The path of motion follows the circular path, whereas the velocity direction is always straight ahead, and so as the force inward is lost, it continues straight on.

When an object with mass is swung around there will be a reactive force as a result of Newton's Third Law (each action has an equal and opposite reaction). This is commonly known as **centrifugal** force. However, it is not a real force and so although in common language, do not use this term. When you go around a corner in a car you experience this outward force as a reaction to the centripetal force, and it is equal in size to the centripetal force, but in the opposite direction (this phenomena is explored at A-level).

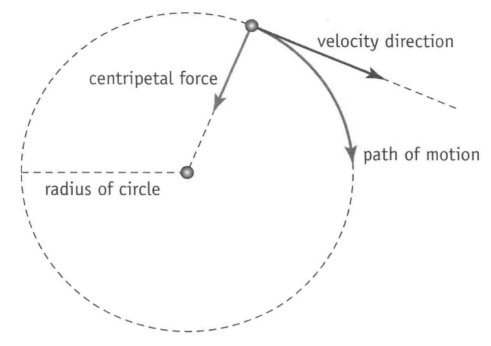

Figure 28 - Relating the velocity, path of motion, centripetal force and radius of the circular movement

More centripetal force is required to keep an object on a circular path when:

1) The **mass** of the object is **increased**.
2) The **speed** of the object is **increased**.
3) The **radius** of the circle is **reduced**.

At any given point of the path of motion, the force inwards is at right angles to the velocity vector. If the mass of an object is doubled, the force required to keep it in a circular motion also doubles. Since the force is inwards, there must be acceleration in that direction also (Newton's second law).

During the circular motion, the direction of the object changes, but the speed does not change. Thus there is a change in its velocity.

4.7 Mechanics - SUVAT

In the OCR specification, a more advanced version of the acceleration topic is discussed. The purpose of this discussion is to relate the different parts of the equations that relate velocity, acceleration, time, and distance.

Acceleration, as we saw above, is the rate at which velocity is changed. So to calculate acceleration we need to know the starting velocity, u, and the end velocity, v. We also need to know how long it took to change between the two velocities. In the discussion that follows we will assume that the rate of change of velocity, or the acceleration, is constant. In other words, if we were pressing the accelerator pedal on a car to increase our velocity, the rate at which the velocity increases is the same each second.

This gave us the equation:

$$a = \frac{v - u}{t}$$

Quantity	Name	Units
a	Acceleration	m/s^2
v	Final velocity	m/s
u	Initial velocity	m/s
t	Time taken to change the velocity between u and v	s

We can also work out the displacement achieved over that time by calculating the average velocity (assuming the change in velocity is constant). Firstly, the average velocity is calculated by adding the two velocities together and dividing by 2.

$$average\ velocity = \frac{v + u}{2}$$

Once we have that value we can use the relationship:

$$velocity = \frac{displacement}{time}$$

Using the standard letters for this equation, and rearranging it to make the displacement the subject, we end up with:

$$s = v_{av} * t$$

Quantity	Name	Units
v_{av}	Average velocity	m/s
s	Displacement at the end	m
t	Time taken to change the velocity	s

Then, using some more substitution using the average velocity equation, from above, for the v_{av} in the displacement equation, we end up with:

	Quantity	Name	Units
$$s = \frac{(v+u)}{2} * t$$	s	Displacement	m
	v	Final velocity	m/s
	u	Initial velocity	m/s
	t	Time taken to change the velocity between u and v	s

Graphically speaking we have calculated the area underneath a straight line, describing the change in velocity over time. As we discussed above, the area underneath that line is the displacement. In the graph, below, the initial and final velocities, as well as the time taken are used to calculated the area, using the normal "area of triangle" formula:

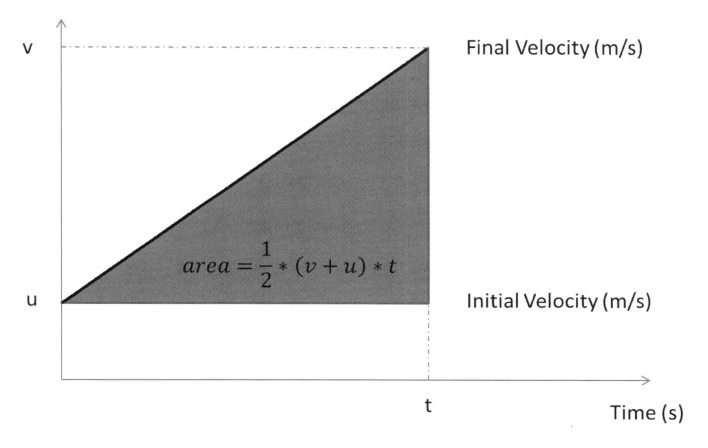

Figure 29 - Area under a graph showing the change in displacement between two velocities.

In the above graph we know that the gradient of the line is the rate of acceleration. So, if we want to calculate the final velocity from the initial velocityand have the gradient, we use the mathematicical concept of a linear equation, where y=mx+c. In this graph the intercept, c, is the initial velocity, u, the gradient, m, is the acceleration and the x-axis is the time. We therefore end up with:

$$v = u + at$$

Quantity	Name	Units
a	acceleration	m/s²
v	Final velocity	m/s
u	Initial velocity	m/s
t	Time taken to change the velocity between u and v	s

So far we have looked at equations that include a simple relationship using time, acceleration, initial and final velocities (and, by derivition, the average velocity). Let's say that we do not know the final velocity, but still want to work out what distance is travelled when we have an initial velocity and have a constant acceleration, a.

We can revisit the displacement equation above and combine it with the velovity equation.

Substituting v in s=1/2 (v+u)*t with v=u+at, we get:

$$s = \frac{1}{2} * \big((u + at) + u\big) * t$$

Simplifying this by multiplying in the ½ into the brackets:

$$s = \left(u + \frac{1}{2} * at\right) * t$$

Then multiplying in the t into the bracket we end up with:

$$s = ut + \frac{1}{2} * at^2$$

Quantity	Name	Units
a	acceleration	m/s²
s	displacement	m
u	Initial velocity	m/s
t	Time taken to change the velocity between u and v	s

Finally, if we do not know the time involved in the motion, but we do know the initial and final velocities, the acceleration and the displacement that occurred, we can connect these with a single equation by substituting the time in v=u+at (t=(v-u)/a) into s=1/2*(u+v) * t above.

$$s = \frac{1}{2} * \frac{(u+v)(v-u)}{a}$$

$$2as = (u+v)(v-u)$$

$$2as = uv - u^2 + v^2 - uv = v^2 - u^2$$

Rearranging to make v² the subject we get the last of the SUVAT equations:

$$v^2 = u^2 + 2as$$

Quantity	Name	Units
a	acceleration	m/s²
s	displacement	m
u	Initial velocity	m/s
v	Final velocity	m/s

We have derived these equations because the all answer a particular question well. The way you use them is not immediately obvious. Below, I outline the method that I have found particularly successful.

Step 1: List what you know, using the standard notation – ensure that the units used in them are base units (see above) – as a list, like the one below (if you do not either need to calculate or use a value, put an 'X' next to it:

S =
U =
V =
A =
T =

Step 2: Put a big question mark (?) next to the physical quantity you are looking to find
Step 3: Choose the equation that has the physical quantities you both have values for AND have the unknown variable
Step 4: Rearrange the equation so that the subject of the equation is the one you identified in step 2
Step 5: Substitute in the values and calculate the answer

Worked example

A particle is accelerated uniformly from rest, so that after 10 seconds it has achieved a speed of 15 m/s. Find its acceleration and the distance it has covered?

S= ?
U= rest, so = 0 m/s
V= 15 m/s
A= ?
T= 10s

This is a two part question. Let's first solve it for the acceleration:

$$a = \frac{v - u}{t}$$

Substitute in the values directly:

$$a = \frac{15 - 0}{10} = \frac{15}{10} = 1.5 m/s/s$$

Go back and enter this in your list so you get

S= ?
U= rest, so = 0 m/s
V= 15 m/s
A= **1.5 m/s²**
T= 10s

Now we can choose an equation that allows us to use these values. There are several that are available, but the simplest is:

$$s = \frac{(v + u)}{2} * t$$

We can substitute in the values directly without rearranging:

$$s = \frac{(15 + 0)}{2} * 10 = 75m$$

Go back again and enter this in your list so you get:

S= **75m**
U= rest, so = 0 m/s
V= 15 m/s
A= **1.5 m/s²**
T= 10s

In the next example we have some a bit more work to do

Worked example

A train travels along a straight piece of track between 2 stations A and B. The train starts from rest at A and accelerates at 1.25m/s² until it reaches a speed of 20m/s. It then travels at this speed for a distance of 1560m and then decelerates at 2m/s² to come to rest at B. Find:

(a) Distance from A to B
(b) Total time taken for the journey
(c) Average speed for the journey

We need to split this into three pieces. The first part is the displacement during the acceleration at 1.25m/s², then the distance as a the constant speed, and then the displacement during the deceleration.

Acceleration phase

S = ?
U = rest so 0
V = 20m/s
A = 1.25m/s²
T = ?

Firstly we choose to solve it for the time taken

$$a = \frac{v - u}{t}$$

Rearranging this to make t the subject and substitute in the values, we get:

$$t = \frac{v - u}{a} = \frac{20 - 0}{1.25} = \frac{20}{1.25} = 16s$$

S = ?
U = rest so 0
V = 20m/s
A = 1.25m/s²
T = **16s**

Now we can work out the displacement during the acceleration using:

$$s = \frac{(v + u)}{2} * t$$

Substituting in the values for the velocities and the calculated time we get:

$$s = \frac{(20 + 0)}{2} * 16 = 160m$$

S = **260m** during the acceleration phase
U = rest so 0

V = 20m/s
A = 1.25m/s²
T = **16s**

Phase two is a linear motion, in other words there is no acceleration occuring. We know that the distance during that phase is 1560m, and the the train is travelling at 20m/s. To calculate the time it took to travel this distance we use:

$$t = d/v$$

we can substitute in the values from above and we get:

$$t = \frac{1560}{20} = 78s$$

The train thus travels at 20m/s over 1560m for 78s.

The final phase has the train decelerating from 20m/s to rest at 2m/s². Since we have a change in velocity; an acceleration, we use SUVAT:

S = ?
U = 20m/s (we start at a higher velocity since we are slowing down)
V = rest, so 0
A = -2m/s² (we are decelerating so the acceleration is negative)
T = ?

Firstly we can solve for time, as above:

$$a = \frac{v-u}{t}$$

Rearranging this to make t the subject and substitute in the values, we get:

$$t = \frac{v-u}{a} = \frac{0-20}{-2.0} = \frac{-20}{-2.0} = 10s$$

S = ?
U = 20m/s (we start at a higher velocity since we are slowing down)
V = rest, so 0
A = -2m/s² (we are decelerating so the acceleration is negative)
T = **10s**

Now we can solve for the displacement:

$$s = \frac{(v+u)}{2} * t$$

Substituting in the values for the velocities and the calculated time we get:

$$s = \frac{(0+20)}{2} * 10 = 100m$$

S = **100m**
U = 20m/s (we start at a higher velocity since we are slowing down)

V = rest, so 0
A = -2m/s² (we are decelerating so the acceleration is negative)
T = **10s**

   a) The answer to the distance between station A and B is thus the total of the distances travelled in the three phases: 160m+1560m+100m =1820m
   b) The total time taken is 16s+78s+10s=104s
   c) The average speed is the total distance travelled (1820m) divided by the total time taken:

$$v_{av} = \frac{1820}{104} = 17.5 \, m/s$$

4.8 Questions

1. A car accelerates uniformly from rest and after 12 seconds has covered 40m. What are its acceleration and its final velocity?
2. A train is uniformly retarded from 35m/s to 21m/s over a distance of 350m. Calculate the retardation and the time taken to come to rest from the 35m/s.
3. A particle is accelerated from 1m/s to 5m/s over a distance of 15m. Find the acceleration and the time taken.
4. A car accelerates uniformly from 5m/s to 15m/s taking 7.5 seconds. How far did it travel during this period.
5. A particle moves with uniform acceleration 0.5m/s2 in a horizontal line ABC. The speed of the particle at C is 80m/s and the times taken from A to B and from B to C are 40 and 30 seconds respectively. Calculate
   a. Speed at A
   b. Distance BC
6. An object has in initial velocity 5m/s, a final velocity of 36km/hr and acceleration of 1.25m/s/s, how far has it travelled once it reached the final velocity?
7. A car accelerates from rest with acceleration 0.8m/s2 for 5 seconds. Find the final velocity
8. A train starts from rest and accelerates uniformly at 1.5m/s2 until it attains a speed of 30m/s. Find the time taken and the distance travelled.
9. A car is being driven along a road at 25m/s when the driver suddenly notices that there is a fallen tree blocking the road 65m ahead. The driver immediately applies the brakes giving the car a constant retardation of 5m/s2. How far in front of the tree does the car come to rest?
10. In travelling the 70cm along a rifle barrel, a bullet uniformly accelerates from rest to a velocity of 210m/s. Find the acceleration involved and the time taken for which the bullet is in the barrel.

Chapter 5. Turning Forces - Moments

Have you ever tried to open a door by pushing it next to its hinges? How does this compare with pushing at the door handle? What you are doing is using a force to turn the door around its hinges. The place an item swings or rotates around is called its **pivot** or **fulcrum**. When a **force** acts on something which has a turning point of **pivot**, a **turning force** called a **moment** is produced. The harder you push, the stronger the moment, and, as you will note from trying to open the door, the longer the distance from the pivot, the stronger the moment.

$$Moment \propto Force$$

$$Moment \propto Distance\ to\ pivot$$

The equation to calculate the turning force (or moment) is only based on these two physical quantities and is thus expressed as:

	Quantity	Name	Units
$M_0 = F \times d$	M_0	Moment (or turning force)	Nm
	F	Force at right angles to the distance	N
	d	Shortest distance to pivot (perpendicular to the force)	m

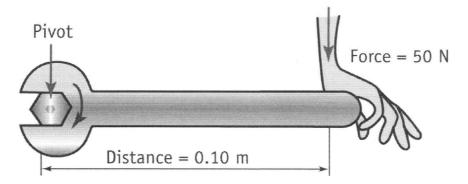

Figure 30 - Using a moment to loosen a nut

In the example of the spanner above, the turning force is 0.10m x 50N = 5.0Nm. If the turning force required to loosen the bolt is less than this 5.0Nm, the bolt will come loose.

Examples of Moments in action
1) Using a door handle to push a door closed.
2) See-saws.
3) Loosening a nut with a spanner.

4) Cranes.
5) Beam-balances.
6) Handle bars on bikes.
7) Levers.

Chapter 6. Moments in balance

For a system to be balanced (in equilibrium), the total anti-clockwise moment must equal the total clockwise moment. This is called the **principle of moments**.

Many of the questions on moments in exam papers use the idea of putting different loads on a seesaw. The mathematics is the same as for the spanner, but the situation is slightly different.

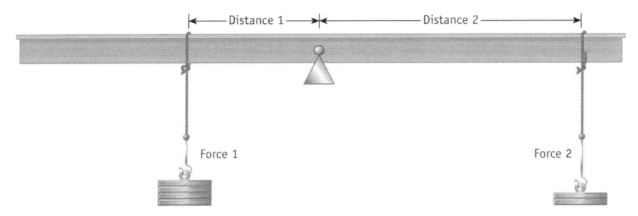

Figure 31 - Principle of moments

For the two weights above to be in equilibrium (so that the beam does not start turning in either direction), the moment counter clock wise must equal the one clock wise.

So:

$$Distance\ 1 * Force\ 1 = Distance\ 2 * Force\ 2$$

If either product was bigger, it would start turning in the direction of the larger force.

Worked example

Looking at the seesaw below, work out the force required to balance it:

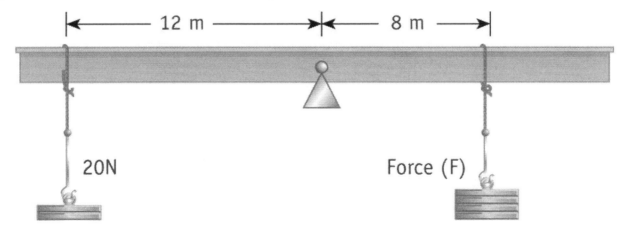

Figure 32 - Work out the Force, F

$d_1 = 12m$
$F_1 = 20N$
$d_2 = 8m$
$F_2 = ?$

$$F_1 * d_1 = F_2 * d_2$$

Rearranging to make F_2 the subject.

$$F_2 = \frac{F_1 * d_1}{d_2} = \frac{12 * 20}{8} = 30N$$

Since forces add, if there were two forces acting on one side at different distances, the total turning force on that side would be the sum of the products of the two turning forces. For example if two forces, F_1 and F_2 acted at distances d_1 and d_2 the total moment for those two forces would be:

$$Total\ moment = F_1 * d_1 + F_2 * d_2$$

It is therefore possible to have multiple loads on one side acting as one combined moment by adding the individual moments together.

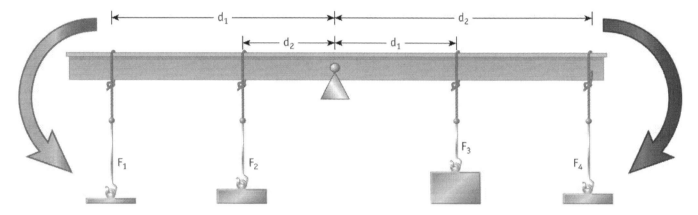

Figure 33 - Multiple moments in equilibrium

In order for these to be in equilibrium, according to the principle of moments the left hand moment, acting counter clock wise (the blue arrow), must be equal to the right hand moment, acting clock wise (the red arrow).

In this case this equation has to be satisfied:

$$F_1 * d_1 + F_2 * d_2 = F_3 * d_3 + F_4 * d_4$$

6.2 Questions

1. A force of 200N is applied to a spanner of length 20cm. Calculate the moment.
2. A uniform metre of mass 100g balances at the 40cm mark when a mass **X** is placed at the 10cm mark. What is the value of **X**?
3. The diagram shows a simple machine for lifting milk bottles

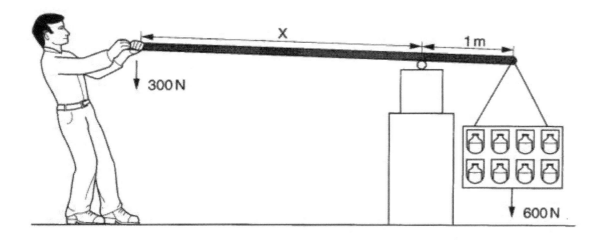

Figure 34 - Work out the distance, X

4. Calculate the turning force (moment) of the pail of milk.
5. Calculate the distance, X, required to lift it with a force of 300N.
6. Explain the following:
   a. A mechanic would use a long spanner to undo a tight nut.
   b. A door handle is far from the hinge.
   c. It is difficult to steer a bike by gripping the centre of the handlebars.
7. The seesaw below is at equilibrium. Calculate the missing distance d?

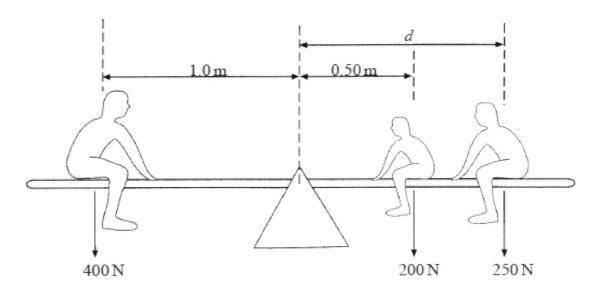

Figure 35 - Work out the missing distance, d

Chapter 7. Centre of mass and equilibrium

The concept of centre of mass (or centre of gravity) is easily misunderstood. All objects have a centre of gravity and it the weight of that object can be considered to reside in one single spot, its centre of mass. At this point the object would be at equilibrium with the same turning forces acting to cancel each other.

An object is a equilibrium when:

1) There is no resultant force
2) There is no resultant turning effect.

Finding The Centre Of Mass Of A Plane Lamina.

Make three holes near the edge of the lamina so that the lamina swings freely when pivoted from each hole. Hang the lamina about one of its holes on a pin clamped on a retort stand.

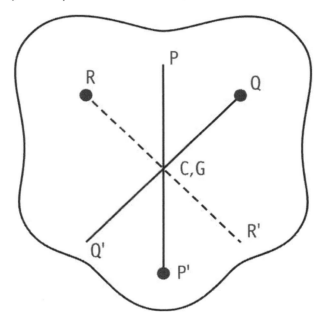

Figure 36 - Centre of mass of an irregular body is where all the lines of action meet

Suspend a plumb line from P and mark the position P' on the lamina.

Repeat the experiment by suspending the lamina from Q and R and similarly mark the plumb line positions Q' and R'. You will observe that all the three lines, PP', QQ' and RR' intersect at one point. This point of intersection of these lines is the centre of gravity of the lamina.

The reason for this is that the centre of gravity, or centre of mass, will always seek to be between the point of suspension and the cause of the gravitational attractsion, e.g. the Earth. In other words, it will hang "straight down".

More regular objects are easier to identify the location of the centre of mass:

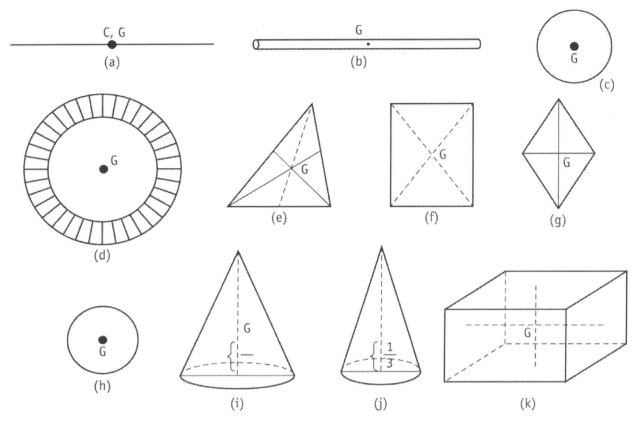

Figure 37 - Centre of mass of regular objects

Stability

Objects are often designed so that they are stable. If an object is pushed and released it is stable if it goes back to its original position. Some objects are more stable than others. Stability can be increased by widening the object's base:

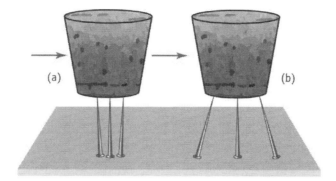

Figure 38 - Stability is increased by widening the base

Or it can be improved by lowering the Centre of Mass. In the example below, the cylinders' Centre of Mass is lowered by adding more mass at the bottom.

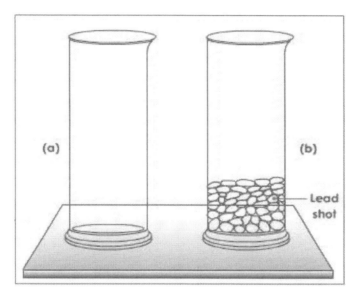

Figure 39 - By adding mass at the bottom of the cylinder, the Centre of Mass is lowered and it becomes more stable

States of equilibrium.

There are three types of equilibrium. We will examine the three by considering a Bunsen burner in three different positions:

1) Standing up, as normal. Place a Bunsen burner on its broad base (figure below). Push the top to one side and see what happens. You will notice that the burner does not fall off unless it is given a hard push. This is because the body is in **stable** equilibrium; it has a broad base, a heavy bottom, thus lowering its centre of mass. When the burner is tilted more and more, the Centre of Mass gets raised and the burner falls back to make the Centre of Mass as low as possible [Figure (a) below]

2) Place the burner upside down as shown in figure below. A slight push causes the Centre of Mass to be lowered and the burner begins to fall to make the Centre of Mass as low as possible [Figure (b) below)]. This makes it **unstable**
3) Let the burner lie on its side as in figure below. Push it slightly and see what happens. On further pushing, the Centre of Mass neither gets raised nor lowered. The burner just rolls maintaining its Centre of Mass at the same level. Objects like cylinders and cones lying on their side roll because they are in **neutral equilibrium** [Figure (c) below)].

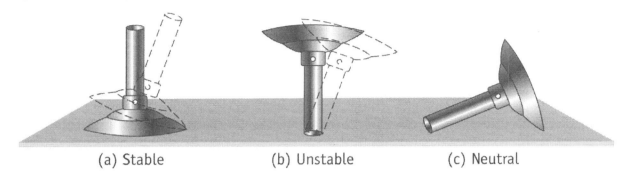

(a) Stable   (b) Unstable   (c) Neutral

Figure 40 - The states of equilibrium

Moving objects need to be designed so that they do not end up unstable. The bus will show how this can be. When a centripetal force is applied to it, it turns inwards, and a reaction to this is that there is an outwards force caused by the mass of the bus. As a result the bus will lean outwards and when it leans too much it will fall over. Understanding the Centre of Mass is essential in the design of the bus.

In position (a), the Centre of Gravity **(G)** is in a stable position. During (b) the bus is turning, but not excessively so. Draw a line right down, representing the weight of the bus from G – this is known as the "line of action". If the bus ceases to experience the outward force, G will cause a moment back towards the stable (a) position. Imagine the tyre acts as a pivot. In position (c) G is now beyond the pivoting wheel and so it will fall over to position (d) as the moment caused G being to the right of the pivot.

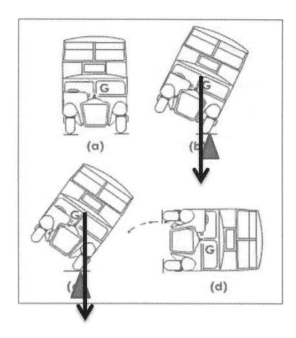

Figure 41 - A bus affected by tilting during a turn

Solving problems with this usually involves finding when the line that represents the force due to weight has passed the point where the object is pivoting.

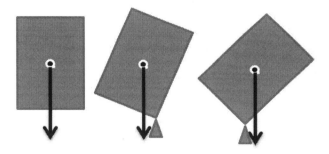

Figure 42 - sequence of blocks with central Centre of gravity

In the blocks to the left, where the Centre of Mass is marked with a circle, the left-most one is in a stable equilibrium, the middle object will fall back to the stable equilibrium as the Centre of Mass is causing a moment in a counter clockwise direction. The one on the far right will cause it to tip over to the right as the moment is then clockwise.

Notice in the next sequence of blocks that the Centre of Mass has been lowered and so the object becomes more stable

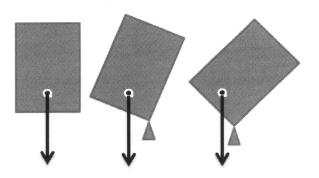

Figure 43 - Sequence of block with a low centre of gravity

The simple method of working these out is to draw a line straight down, called the line of action, and seeing which side of the pivot (edge of the base) it ends up on. It will fall in that direction.

Figure 44 - Formula 1 cars have wide bases and a low centre of mass

7.1 Pendulums

This part of the AQA syllabus is very brief, but introduces you to the concept of an object that oscillates due to a transformation of one type of energy into another, and then back again. This is revisited in most A-level physics syllabi later.

In the case of a pendulum a bob is suspended on a sting and lifted up a bit, while keeping the string tight and then let go. By lifting the bob up it gains gravitational potential energy, which it would like to remove. When it is let go, it accelerates, reaching maximum velocity at the bottom of its swing (the equilibrium position), where it therefore has the highest kinetic energy.

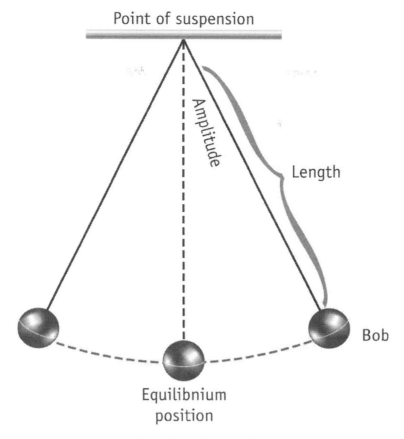

Figure 45 - typical pendulum

Some technical terms are required to dicsuss this movement:

1) The **amplitude** of the oscillation is the distance the bob moves from the equilibrium position to the furthest point in its swing (in this diagram that is either at the extreme right location of the bob, or the extreme right position)
2) The **time period** is the time it takes for the pendulum to move one full oscillation. The bob must end up at extactly the same position, moving in the same direction the timer was started at. This is a point that is often missed. For instance, the bob, if you started measuring it at the equilibrium position, will move to the extreme right, then back to the equilibrium position, then to the extreme left, and then back to the equilibrium position. That means that it was at the equilibriu position three times. But only on the first and the third passing was it going in the same direction.
3) The **frequency** is the inverse of the time, f=1/time period. This equation is revisited later in the waves section of this book. What it describes is the number of oscillations each second and is meansured in Hertz. E.g. a pendulum with a period of 0.1s has a frequency of 1/0.1=10 oscilllations per second, or Hertz (Hz).

There are a number of physical parameters that you can change with regards to the pendulum:

1) The length of the string
2) The mass of the bob at the end of the string
3) The height at which the bob is let go

It might be counter-intuitive, but on the length of the string will change the time period, and therefore the frequency.

7.2 Questions

1) Draw a coke bottle in the 3 states of equilibrium.

2) Comment on the stability of a double decker bus & racing car.

3) Look at the image below of a Fosbury flop and work out why do the high jumpers choose to us it?

Figure 46 - What makes the Fosbury flop so effective?

4) A student performs an experiment using a pendulum. He measures the time taken for 20 oscillations as 40.15s, 41.12s, 45.21s and 40.87s.

    a. Which measurement is anomalous?

    b. Work out the mean of for the 20 oscillations (remember NOT to use the anomalous measurement)

    c. Work out the period of one oscillation

    d. Work out the frequency of the oscillations

5) What is the one way that you can change the time period of a pendulum?

6) What happens to the frequency when the time period becomes smaller?

7) Sketch an energy transfer diagram for the pendulum as it goes through one complete oscillation.

Chapter 8. Density

Density is defined as the amount of mass in a certain volume. Really dense materials contain a lot of mass in small volume. At a particle level, it can relate to how close the particles are to each other and how massive each particle is.

Mathematically it is defined as the mass per unit volume and is expressed as:

$$Density = \frac{Mass}{volume} \text{ or } \rho = \frac{m}{V}$$

Quantity	Name	Units
ρ	Density	kg/m³ or g/cm³
V	Volume	m³ or cm³
m	Mass	kg or g

8.1 Measuring density

To measure the density of an object there are two quantities required:

1) Mass – this can be measured using a mass balance
2) The volume – this can be measured on two ways
   a. For a regular object you would measure the sides and work out the volume using the normal equations. For instance, a regular block you would use the product: length x height x depth
   b. For irregular objects you immerse the object in water and measure how much water has been displaced. Although mainly used for measuring the volume of an irregular object, it can also be used for regular objects, as in Figure 47 - measuring the density of an object using the amount of water displaced..

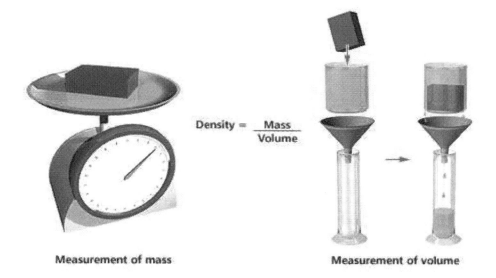

Figure 47 - measuring the density of an object using the amount of water displaced.

In a more common situation at a school lab you will immerse an object in a measuring tube and calculate the change in the measured volume. This will give you the volume of the object:

Worked example

A stone of mass 25g is submerged in a measuring tube that originally read 30cm³. After the submersion, the level of the liquid rises to 40cm³. What is the density of the stone?

Difference = 40 – 30 = 10cm³

$$\rho = \frac{25g}{10cm^3} = 2.5g/cm^3$$

8.2 Questions

1) An object has a mass of 100g and a volume of 20cm³. What is its density?
2) An object has a volume of 3m³ and a density of 6000kg/m³. What is its mass?
3) The density of air is 1.3kg/m³. What mass of air is contained in a room measuring 2.5 x 4 x 10m?
4) Copy and complete the table:

OBJECT	DENSITY (kg.m³)	MASS (kg)	VOLUME (m³)
A		4000	2
B	8000		4
C	2000	1000	
D		2000	4

(a) Which object has the greatest mass?
(b) Which objects would be made of the same substance?
(c) Which object would float on water?

   (density of water = 1000kg/m³ – 1g/cm³).

5) In a stack are 1000 bricks measuring when stacked 2m x 1m x 1m.
   a. Calculate the volume.
   b. Calculate the volume of 1 brick.
   c. If the density of brick is 2500kg/m³, what is the mass of the stack?
   d. When put into a van, the van's max load is 1000kg. How many bricks can he load?

Chapter 9. Pressure

When a force is applied across an area, that force is either spread out over a larger area, as in a person wearing a pair of snow shoes, or it is focused in a small area, as in the tip of a nail.

The pressure is higher when the area is smaller, and it is bigger, the higher the force. It is useful to consider it as the force per unit area:

$$Pressure = \frac{Force\ (N)}{Area\ (m^2)} = \frac{F}{A}$$

Quantity	Name	Units
P	Pressure	Pa (Pascals) or N/m²
F	Force	N
A	Area	m²

Practical examples

Understanding that the area of an object either allows the force to be spread out wider, or be focused in a small area can be used to good effect. It is often desirable to ensure that an object does not sink in mud or a softer substance, or an object pierce another object. Either can be achieved easily by designing the area between the two environments. Pressure is measured in N/m² but is also known as Pascals. 1 N/m²=1Pa.

Wall Foundations	Drawing pin
Provides a large surface area under the supports (footings) to spread the force over a wide area, stopping the house from sinking into the ground	Provides a very small surface area so that the force at the tip focuses the force from the thumb end. This high force causes the pin to enter the wall

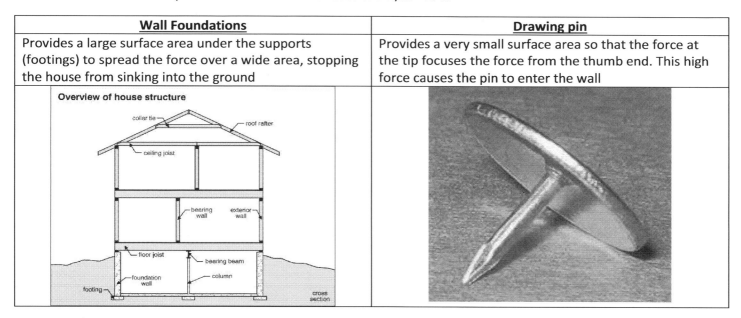

Note that pressure is both measured in Pa and in N/m². Pa is pronounced Pascals in honour of the French physicist Blaise Pascal (19 June 1623 – 19 August 1662) – pictured to the right.

Figure 48 - Blaise Pascal

Worked Examples

1) The wind pressure on a wall is 100Pa. if the wall has an area of 6m², calculate the force.

$$F = P * A$$

$$F = 100Pa * 6m^2 = 600N$$

2) A concrete block has a mass 2600kg (use W=m*g, for this calculation use g=10m/s²). What is the **maximum** pressure it can exert on the ground?

Maximum pressure will be from the side with the SMALLEST area. The dimensions of the block are 2m x 1m x 0.5m. The smallest area is thus 1 x 0.5m = 0.5m²

$$P = \frac{F}{A} = \frac{2600 * 10}{0.5} = 52000 Pa$$

9.2 Questions

1) A force of 200N acts on an area of 4m².
   a. Calculate the pressure
   b. Recalculate for half the area.
2) What force is produced if:
   a. Pressure = 1000Pa, Area = 0.2m².
   b. Pressure = 2kPa, Area = 0.2m².
3) Explain why a load – spreading washer prevents a nut sinking wood.

4) A rectangular block of mass 30kg measures 0.1M by 0.4M by 1.5M.
   a. Calculate the weight of the block.
   b. Calculate the block's maximum pressure on the ground.
   c. Calculate the block's minimum pressure on the ground.

9.3 Pressure in liquids

In liquids, pressure has the following properties:

1) Pressure acts in **all directions**: the liquid pushes on every surface it contacts.
2) Pressure **increases with depth**: due to the weight of the liquid **above** the object.
3) Pressure depends on the **density** of the liquid: viscous liquids (e.g. oil) have high densities.

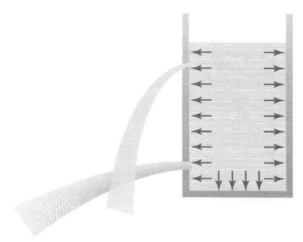

Figure 49 -- Pressure at the bottom is higher than at the top because the weight of the liquid above

This relationship is essential when working below water or when building dams. The lower one dives in water, the higher the pressure on the diver and the taller the dam, the higher the pressure at the bottom of the dam.

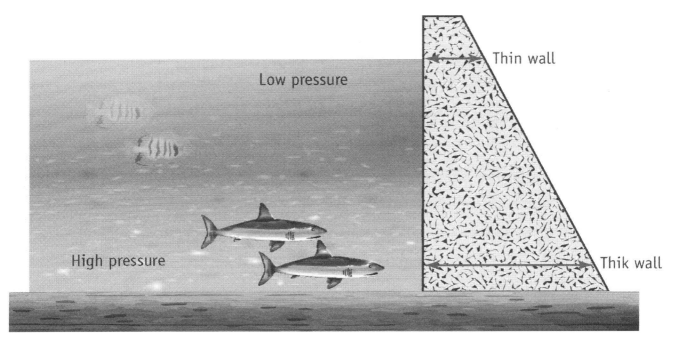

Figure 50 - Dams are built stronger at the bottom to ensure they are strong enough to withstand the additional pressure at the bottom

Calculating pressure in liquids

From the brief discussion above, we can see that the pressure should be directly proportional to weight of the water above the point.

$$Pressure = density * gravitational\ field\ strength * height\ (or\ depth)$$

Using symbols instead:

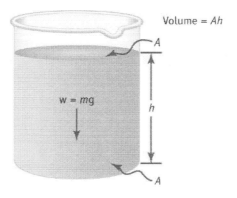

Figure 51 - Pressure at a depth, h, in a cylinder of water

$P = \rho * g * h$	Quantity	Name	Units
	P	Pressure	Pa (Pascals) or N/m²
	$\rho(rho)$	Density	kg/m³
	g	Gravitational friend strength	m/s²
	h	Height (or depth)	m

Worked example
If the density of water is 1000kg/m³, what is the pressure at the bottom of a swimming pool 2m deep?

$$P = \rho * g * h$$
$$P = 1000 * 10 * 2 = 20000 Pa$$

9.4 Questions

g = 10m/s²; Density of water = 1000kg/m³; Density of paraffin = 800kg/m³.

1) The experimental tubes to the right contain water.
a) Compare the pressure at B with A.
b) Compare the pressure at B with D.
c) Compare the pressure at A with C.
d) Calculate the pressure at B due to the water.

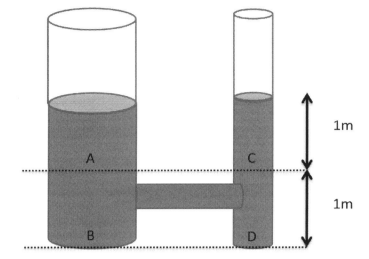

2) A storage tank measures 4m x 3m and is filled with paraffin to a depth of 2m. Calculate:
   a. The volume of paraffin.
   b. The mass of paraffin.
   c. The weight of paraffin.
   d. The pressure due to paraffin at the bottom of the tank.
3) A glass fish tank has a base area of 3m². It is filled with water to a depth of 4m.
   a. Calculate the volume of water.
   b. Calculate the mass of water.
   c. Calculate the weight of the water.
   d. Calculate the pressure at the bottom of the tank.
4) If the density of saltwater is 1714kg/m³, recalculate
   a. Its mass.
   b. Its weight and its pressure on the fish tank above if filled to a depth of 6m.

9.5 Hydraulics

Liquids do not compress

We speak later about the fact that liquids are virtually incompressible. This means that when a force acts on a liquid, the resulting pressure is the same throughout the liquid. Imagine compressing a cycle pump full of air, and with your finger over the outlet. The pump handle will move into the pump and the gas (air) will compress. The pressure in the cycle pump will increase. Imagine doing the same but filling the cycle pump with water. The pump handle will not move into the pump. This illustrates that a liquid does not compress.

As a result, the pressure you put on the liquid will be the same throughout the pump.

When the area of the parts of the pump change

In the following diagram, a force of 50N is exerted on the wider end of a hydraulic pump using a piston (a rod pushing into the pump). The pump is filled with a liquid, such as an oil. The resulting pressure is the same throughout the liquid. As such the pressure at the other end, on the other piston will be the same.

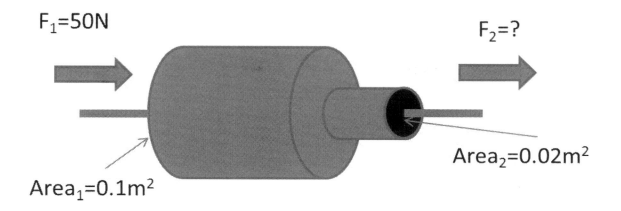

Figure 52 - A hydraulic pump (in reverse)

We can therefore calculate the resulting force pushing outwards.

Pressure in is $P_1=F_1/A_1=50/0.1=500Pa$. That pressure will cause a force outwards at the other end of $F_2=P_2*A_2=500 * 0.02=10N$. This is not very useful, except if you want to move an object a longer way as a result of pushing the piston a short distance.

However, if we reverse the process.

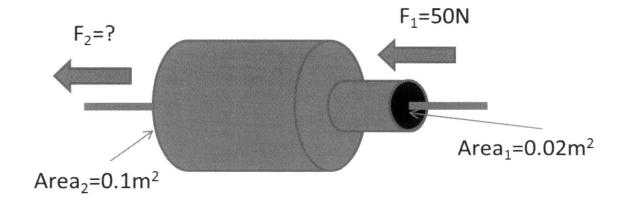

Figure 53 - hydraulic pump

We set the pressures equal to each other, so $P_1=P_2$:

$$P_1 = P_2 \text{ therefore } \frac{F_1}{A_1} = \frac{F_2}{A_2}$$

Rearranging this to find F_2:

$$F_2 = \frac{F_1 A_2}{A_1} = 50 * \frac{0.1}{0.02} = 250N$$

So, by extering a force of 50N at one end, the force is multiplied up and becomes 250N at the other end. This type of configuration is called a hyraulic pump and can be used for:

1) Lifting cars in garages
2) Pressing the brakes in a car
3) Moving the digger scoop on a mechanical digger
4) Moving flaps and lowering/raising the wheels on a plane
5) Power assisted steering in a car or lorry
6) Opening and closing huge vault doors

9.6 Atmospheric pressure

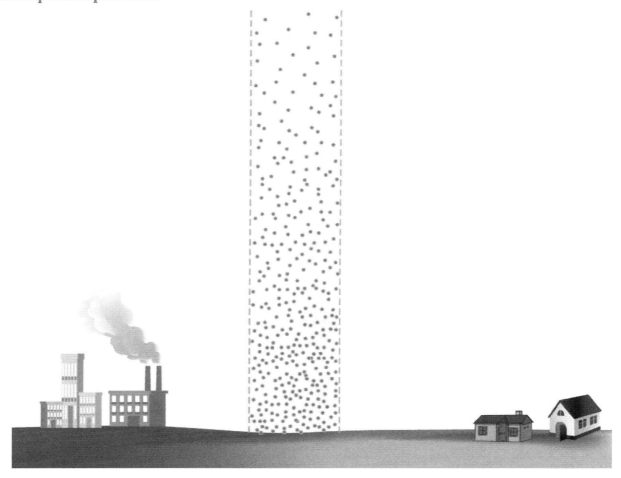

Figure 54 - The density of the air reduces as your go higher. The weight of that air causes the atmospheric pressure at lower levels

We live in an ocean of air, where the air above is exerts pressure down on us due to its weight.

The pressure of the atmosphere has 2 main properties:

1) Its pressure acts in **all directions**.
2) Its pressure becomes **less** as you rise up through it.

The atmosphere is **denser** at lower levels. This is because gases can be **compressed**.

Atmospheric Pressure

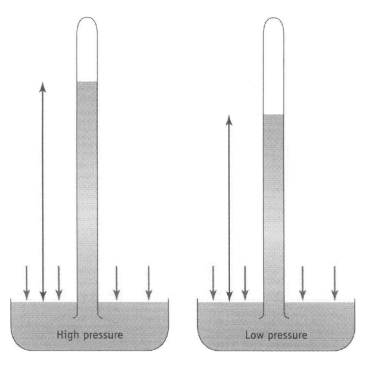

Figure 55 - Mercury Barometer

At sea level, atmospheric pressure is approximately 100,000Pa (100kPa). 100,000pa is known as standard atmospheric pressure or 1 atmosphere. 1 atmosphere is the pressure that will support a column of mercury 760.00mm high. In weather forecasting the unit of pressure is the millibar (mb)(100Pa).

Measuring Atmospheric Pressure

The atmosphere changes pressure due to the weather; movement of winds, temperature and the temperature of the ocean affects it. Depending on the time of year, it is possible to determine if it is going to rain if the pressure changes. The atmospheric pressure is measured using a **mercury barometer**. Mercury column held up by air pressure. Mercury height will vary due to pressure differences caused by the weather.

The column of mercury houses a vacuum at the top. This vacuum is expanded as the pressure from the atmosphere drops since the weight of the mercury is no longer being pushed up by the air pressure. The height of the column has a pressure at the bottom equal the P= ρgh, where ρ is the density of mercury. When the pressure due to its weight is equal to the mercury, the column is said to be supported by the atmospheric pressure. When the pressure changes, the column rises, or falls. By leaning the column over, it is possible show how the vacuum is filled by the mercury up to that height.

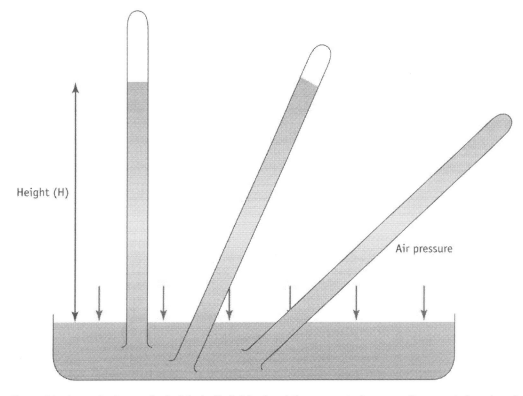

Figure 56 - As a tube leans, the height is diminished and the amount of mercury it supports is reduced

Measuring Pressure Difference

Pressure differences are measured using a mercury manometer. (there are also water and alcohol manometers.) **Manometer** filled with mercury. Height difference shows the extra pressure of the gas supply. This is called excess pressure. To find the actual pressure of the gas supply: atmospheric pressure + excess pressure.

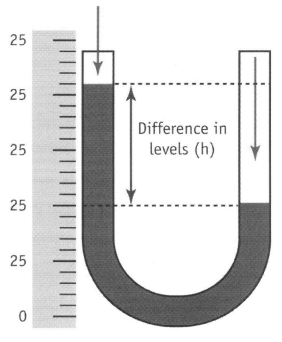

Figure 57 - Manometer

The actual pressure above atmospheric in pascals can be worked out using the formula:

Pressure = depth x density x gravity = 0.11 x 1000 x 10 = 110 Pa

In this example we are using a water manometer, so the density is 1000kg/m^3.

9.7 Questions

1) What is the excess pressure of the gas supply (in mm of mercury) in the diagram above?
    a. What is the actual pressure of the gas supply?
2) If a mercury barometer were carried up a mountain, how would expect the height of the mercury column to change?
3) Explain why you can suck Coca-Cola up through a straw.

Density of mercury = 1350kg/m^3.

4) Calculate the pressure at the bottom of a column of mercury 1.5m long.
5) If on a particular day, atmospheric pressure is 760mm of mercury, what is this in:
    a. Pascals
    b. Atmospheres
    c. Millibars
6) A hydraulic pump with primary end with an area of 0.3m^2 has a piston that is being pressed with a force of 400N. The secondary, lifting end, requires a force of 1200N, what area should that piston have?
7) A brake uses a hydraulic force multiplier. The primary end, with the brake pedal, can be pressed by a human at a maximum of 1200N, the secondary, brake-pad end requires a force of 9000N and has an area of 0.2m^2. What is the maximum area the brake-pedal and piston should have?

Chapter 10. Energy

Energy is essential to all living and to making things occur. Without energy, nothing will change. When the Universe commenced, only a limited supply of energy came into existence.

Any form of energy may be transformed into another. For example, potential energy is converted into kinetic energy when objects are given freedom to move to different position such as when an object falls under the influence of gravity. In all such energy transformation processes, the total energy remains the same, and a transfer of energy from one system to another results in a loss from one to compensate for any gain to another. This is the principle of the conservation of energy which says that energy cannot be created or destroyed.

In the short discussion above, it was highlighted that energy exists in different forms.

10.1 Types of energy

1) Kinetic → energy of motion.
2) Potential → stored energy. Due to an object's position or shape:
    a. Elastic
    b. Gravitational.
3) Heat/thermal → movement energy of atoms.
4) Radiated → energy carried as waves.
5) Nuclear → atoms split releasing energy.
6) Electrical → transfer of energy by electrons.
7) Chemical → energy from fuels.
8) Light or electromagnetic radiation
9) Sound

10.2 Transformation of energy.

Energy is transformed when it **changes** from one form to another. The amount of energy stays the **same**. It is **conserved**.

Principle (or law) of Conservation of Energy

{ *"Energy cannot be created nor destroyed but can only change into different forms."* }

Example

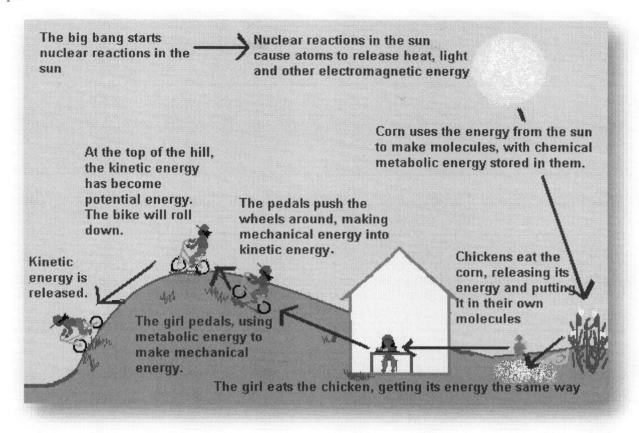

Figure 58 - Example of conservation of energy and energy transfers

This is an example of a rather complex series of energy transfers, starting the Big Bang.

Nuclear -> Chemical in corn -> Chemical in chickens -> Chemical in girl -> kinetic -> Gravitiational potential at the top of the hill -> kinetic energy again

This is an example of an energy transfer (or transformation) diagram.

10.3 Questions:

1) Draw the following energy transformations:
    a. Rocket taking off.
    b. Jack-in-the-box.
    c. Battery in a toy car.
    d. Catapult.
    e. Nuclear explosion.
    f. Gun firing.
    g. Dam.
    h. Firework.

10.4 Energy transfer diagrams (or Sankey diagrams)

It can be useful to have a way of showing how energy in transferred between types using something visual and easy to interpret. Sankey diagrams perform this purpose, showing how much energy is input into a system and how this energy is transformed into different types. All energy transformations are "inefficient" and so some can be considered as "wasted". It is not lost, or destroyed, but transformed into energy that is not useful. These diagrams show where the wasted and useful energy ends up, ensuring that the amount of energy that is input is the same as the sum of the energies released (both useful and wasted). Making sure that the two balance is the same as ensuring that it conforms to the conservation of energy rule.

In the example below, a Sankey diagram shows how much energy a traditional, filament, light bulb is useful and is wasted as heat:

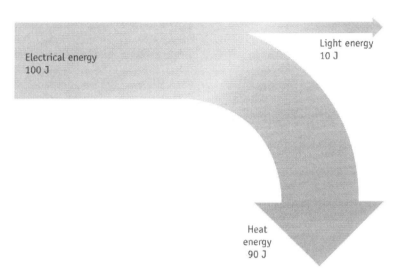

Figure 59 - Snakey Diagram for a filament ligh bulb

Note that 100J is "input" and the output total is 10+90J; also 100J. Only 10J is useful energy, the other 90J is used in heat. A more modern energy efficient bulb shows a similar Sankey diagram, but it is immediately possible to determine that more energy is transformed to useful energy.

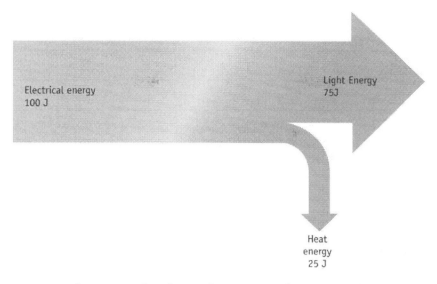

Figure 60 - Sankey diagram for an energy efficient light bulb

Efficiency

Some objects are more EFFICIENT at energy conservations. But none are 100% efficient! Efficiency compares how much of the input energy is transformed to useful energy.

$$Efficiency = \frac{Useful\ Energy\ Output}{Total\ Energy\ Input} \times 100\%$$

Quantity	Name	Units
Efficiency	How efficient the system is	%
Useful Energy Output	How much energy of the total is transformed into useful energy	Joules (J)
Total energy Input	The amount of energy that is provided to the system	Joules (J)

In the filament bulb example above:

For every 100J of energy input to the bulb, only 10J are output as useful light energy:

Efficiency = 10/100 x 100 = 10%

10.5 Questions

1) Calculate the efficiency of a bell using 250J and losing 27J as heat energy.

2) Calculate the useful energy output of a motor 85% efficient and a total energy input of 500J.
3) Calculate the total energy input of a device with an efficiency of 72% and a useful energy output of 5000J.
4) An engine does 1500J of useful work with each 5000J of energy supplied to it. What is its efficiency? What happens to the rest of the energy supplied?

10.6 Calculating Potential, Elastic and Kinetic Energy

Potential energy can be transformed to and from kinetic energy in many ways. For example, a pendulum does it constantly, and repeatedly:

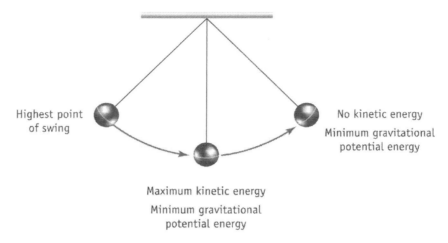

Figure 61 - Energy transfers in a pendulum

When the bob is at its highest point, it has its maximum gravitational potential energy. As it falls, it gains kinetic energy until that is at its maximum at the bottom, when it was transferred all its gravitational energy to kinetic (movement) energy. It then, gradually, converts that kinetic energy back to gravitational potential energy when it rises up to the other extreme. Eventually the pendulum will stop as a result of it transforming some of the energy to heat in friction with the air.

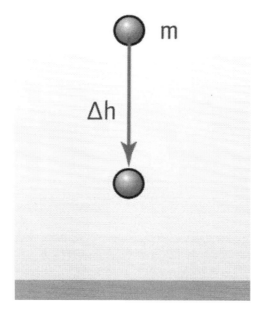

Figure 62 - G.P.E. is relative to the change in height.

Calculating gravitational potential energy (G.P.E.)
The amount of potential energy stored in an object is determined by its height within the gravity of an object. The stronger the gravity, the higher an object and the more massive an object, the more the G.P.E. that object stores.

$E_p = mg\Delta h$	Quantity	Name	Units
	E_p	G.P.E	Joules (J)
	m	Mass	Kg
	g	Gravitational field strength	m/s²
	Δh	Vertical height difference	m

Example: if a 2kg mass is 3M above the ground and gravity is 10m/s², calculate its P.E. (note, in some GCSE specifications g is 9.8m/s² or even 9.81m/s² – in an exam you will be given the number to use).

E_p = 2 kg x 10m/s² x 3N = **60J**.

Calculating stored elastic potential energy
When a spring is extended, work is done on it and this is stored in the spring. The energy stored for a particular extension is:

$$E_e = \frac{1}{2}ke^2$$

Quantity	Name	Units
E_e	E.P.E	Joules (J)
k	Spring constant	N/m
e	Extension of spring	m

10.7 Calculating kinetic energy (K.E.)

When something is moving it has energy. Imagine trying to stop a run-away horse or a rugby forward moving at speed. This type of energy is called the KINETIC ENERGY of the object. There will be more energy 'locked away' in a massive lorry that is moving at 20 m/s than in a motorbike travelling at the same speed and the faster it goes the more kinetic energy it has. The kinetic energy of an object therefore depends on two things:

1) the mass of the object (m)
2) its speed (v)

$$E_k = \frac{1}{2}mv^2$$

Quantity	Name	Units
E_k	K.E.	Joules (J)
m	Mass	Kg
v	Velocity	m/s

Example:
If a 10kg mass has a velocity of 5m/s calculate its K.E.

$$E_k = \frac{1}{2}mv^2 = \frac{1}{2} * 10 * 5^2 = 125J$$

10.8 Questions

1) An object has mass 6kg. what is its G.P.E.
    a. 4m above the ground
    b. 6m above the ground
2) An object of mas 10kg has a speed of 15m/s.
    a. What is its K.E.?
    b. Recalculate its K.E. if its speed is doubled.
3) A stone is dropped from 2.5 km high. Its mass is 200g. Calculate its G.P.E.
4) An object of mass 24kg is released from a height of 45M.
    a. Calculate its G.P.E.
    b. Recalculate its G.P.E. when its mass is doubled.
    c. Recalculate its G.P.E. when its mass is halved.
    d. Recalculate its G.P.E. when its mass is tripled.
5) A car of mass 1000kg travels at 30m/s.

a. Calculate its K.E.
b. It slows to 10m/s. Calculate its K.E.
c. Calculate the change in K.E.

10.9 K.E. and G.P.E Problems.

Worked example 1 – calculating loss of E_p
If the stone on the right is dropped, what is its K.E. half way down?

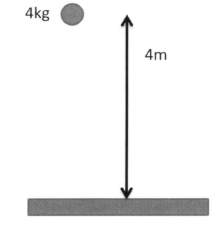

The stone's loss in G.P.E. is equal to its gain in K.E. (Remember energy is conserved)

Therefore: height lost by stone = 2m.

G.P.E. Lost by stone = 4kg x 10 x 2 = 80J.

so, K.E. Gained by stone = 80J.

Worked example 2 – calculating resulting velocity
A stone slides down a smooth slope, as in the diagram. What is its velocity when it reaches the bottom?

Again P.E. at the top = K.E. at bottom.

Therefore: E_p = 4kg x 10 x 5 = 200J.

So E_k at bottom on slope = 200J

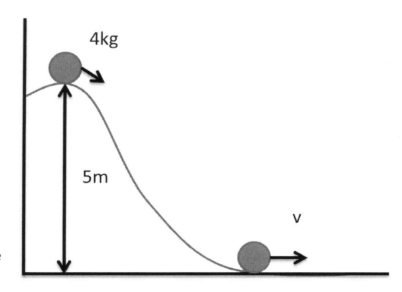

:   So → ½ mv² = 200J
:   So → 2 x v² = 200J
:   So →   v² = 200/2
:   So →   v² = 100
:   So →   v = 10m/s.

This is the long way to work it out. As we saw in the section in weight, all objects accelerate at the same rate due to their weight. It does NOT depend on their mass. We can use some simple maths to derive an equation that will calculate the velocity in one simple step. Learn how to derive this and save time in an exam:

$$E_k = E_p \text{ when fully converted so } \frac{1}{2} * mv^2 = mgh.$$

First simplify by cancelling the mass (m) from both sides, and make v² the subject.

$$v^2 = 2gh, \text{ and so taking the root, } v = \sqrt{2gh}$$

Doing that long calculation again we have

$$v = \sqrt{2 * 10 * 5} = \sqrt{100} = 10 m/s$$

10.10 Questions:

1) A ball is at the top of this slope
   a. Calculate P.E. compared to that at the bottom
   b. Calculate velocity at bottom of slope.
2) A ball of mass 3kg is dropped from the Leaning Tower of Pisa. Calculate its velocity when it hits the ground. (58m).
3) A pole-vaulter has a mass of 50kg.
   a. What is her weight in Newtons?
   b. If she vaults to 4M high, what is her P.E. at the top?
   c. How much K.E. does she have just before reaching the ground?
   d. What is the velocity?
4) Draw the following table:

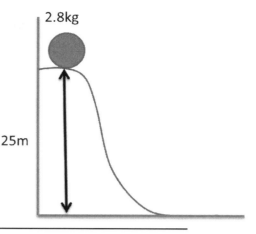

HEIGHT(m)	P.E.(J)	K.E.(J)	VELOCITY(m/s)
0.1			
0.2			
0.3			
0.4			
0.5			
0.6			

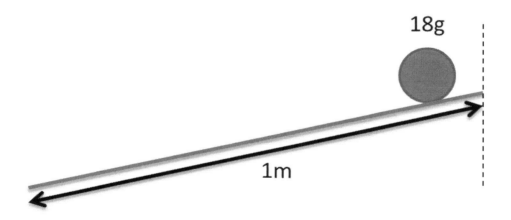

(Assume no friction or air resistance)

a) Calculate P.E. & K.E. for the ball.
b) Calculate its velocity at the bottom,
c) Draw a line graph for any of the following:
   1. Height vs P.E.

2. Height vs K.E.
3. Height vs Velocity.
4. P.E. vs K.E.
5. P.E. vs Velocity
6. K.E. vs Velocity.

Worked example 3 – calculating height

A stone of mass 2 kg is dropped from height H. its velocity is 10m/s. Calculate:

a) The kinetic energy.
b) The drop height

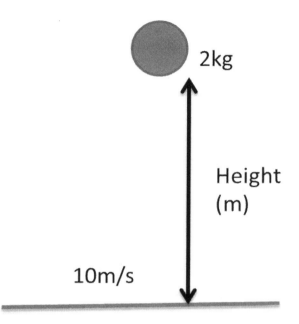

K.E. = ½ mv²

So → 1kg x 10² = K.E.

So → K.E. = 100J.

So → K.E. at bottom = P.E. at top.

So → P.E. = 100J

So → mgh = 100J

h = 100/Kg = 100/20 = 5m

For the more mathematically able, use this equation from above:

$$v^2 = 2gh$$

Rearranging to make h the subject we get

$$\frac{v^2}{2g} = h \text{ substituting in the values } \frac{100}{2*10} = 5m$$

Notice that you do not need the mass in order to calculate the velocity or the height in these questions. All objects accelerate at the same rate in a gravitational field.

10.11 Questions

1) The P.E. of a bullet of mass 50g at the height of its arc is 250J. Calculate the height.
2) An object of mass 15 kg is dropped from height h. Its K.E. at the bottom is 2000J. Calculate the drop height.
3) A firework of mass 3kg produces 450J of K.E. when it hits the ground. Calculate its maximum height.
4) A 2kg ball is thrown vertically upwards at a velocity of 8m/s. Calculate the height reached.

Chapter 11. Energy resources

Energy for electricity is obtained from chemical energy in fuels. Fuels come from non-renewable energy sources e.g. coal, oil and natural gas. All initial energy comes from nuclear fusion in the sun (atoms combine to release energy).

11.1 Electricity generation

Electricity is produced by spinning turbines and generators in power stations. The energy from fuels used to heat water and produce high pressure steam, which is then used to spin turbines at high speeds and turn a generator producing electricity (you will find out later how this is done in more detail).

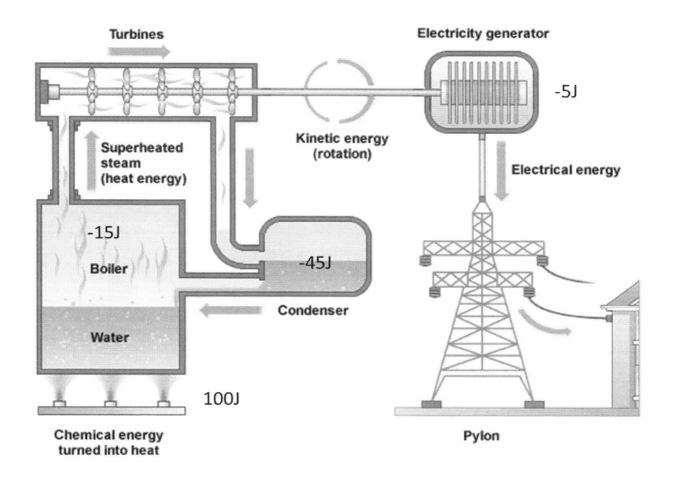

Figure 63 - Energy transformation in the production of electricity

In this process for every 100J the power station burns in fuels, 15J is lost in the boiler inefficiency, 45J is lost in the waste water and a further 5J is lost in the generator. Adding up all the lost energy (15+45+5=65J) we find only 35J is useful output energy.

11.2 Non-Renewable fuels

Most power stations use **non-renewable** energy sources i.e. coal, oil, natural gas and nuclear fuels (Uranium and Plutonium).

All have advantages and disadvantages:

FUEL	ADVANTAGE	DISADVANTAGE
COAL	<ul><li>Lots of heat.</li><li>Reserves left.</li><li>Easy to transport.</li><li>High energy content.</li><li>Cheap to buy.</li></ul>	<ul><li>Finite.</li><li>Expensive to mine.</li><li>Emits CO_2 →Causes global warming.</li><li>Emits SO_2 →Causes acid rain.</li><li>"Dirty Fuel."</li></ul>
OIL	<ul><li>High energy content.</li><li>Numerous applications e.g. plastics, fractions.</li></ul>	<ul><li>Finite.</li><li>Expensive to mine.</li><li>Polluting as coal.</li></ul>
GAS	<ul><li>High energy content.</li><li>"Clean" fuel.</li></ul>	<ul><li>Finite.</li><li>Difficult to store.</li><li>Polluting.</li></ul>
NUCLEAR e.g. Uranium	<ul><li>Huge energy production.</li><li>"Clean" fuel.</li></ul>	<ul><li>High safety methods.</li><li>Power stations are expensive to build</li><li>Risk of radiation</li><li>Difficult to store once depleted</li></ul>

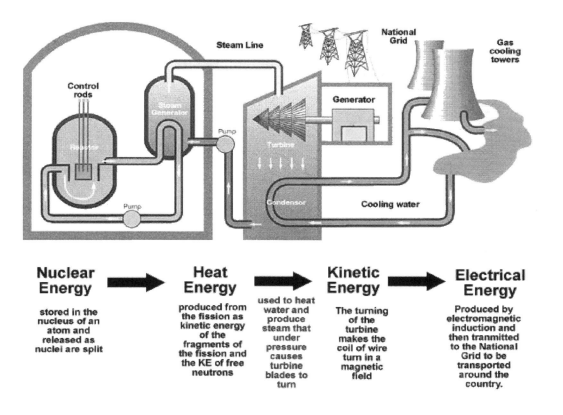

Figure 64 - Nuclear Power plant. The initial heat to the steam generator is carried in a CO_2 gas so that the nuclear fuel does not corrode from water.

11.3 Electricity from water

Water is comparatively dense and is moved around the Earth in two ways that can easily be used to produce electricity.

Tidal power

Firstly, the water is moved by tidal power, caused by the Moon and (to a lesser degree) by the Sun. As the Earth rotates, the water is lifted higher. This can be used by allowing it to flood over a dam built over an estuary, trapping the water behind the dam. When the water subsides, the water inside the dam is higher than the outside and gates are opened to allow the water to flow through turbines back into the sea. The energy transfer diagram would look like this:

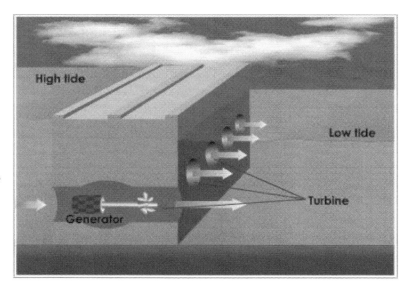

Gravitational energy from the Moon -> E_p -> E_e -> Electricity

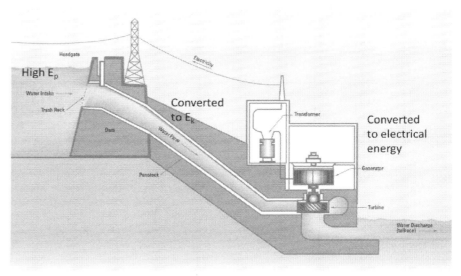

Figure 65 - Tidal power dam

Hydroelectric power

Secondly, the water is heated in the sea by the sun, turned into steam through evaporation from the surface. This is then blown over land to high areas over mountains where it is released as rain. This is moved down through rivers. A dam is built across a large such river, creating a large, artificial lake behind the dam that stores the water at a high E_p. By opening a water intake the water is allowed to drop, gaining E_k through pipes until it reaches a turbine and is converted to electricity. The energy transfer would look like this:

Figure 66 - Hydroelectric Dam

Heat energy from the Sun -> E_p -> E_e -> Electricity

Figure 67 - The Hoover Hydroelectric power station and dam in the USA

11.4 Electricity from air

Wind farms and Areogenerators

As the sun heats the air, it causes it to move as wind. This movement carries kinetic energy and using blades on aerogenerators this can cause a spinning motion in a generator that transfers into electrical energy. Wind farms are controversial because most are very inefficient and might not be as green as initially thought. They are also very unsightly and can cause damage to wild life, in particular birds of prey that mistake the sound of the blades for the movement of birds. Wind farms on land are usually built in areas of natural beauty, sadly destroying the views for generations to come. Wind farms built at sea are usually much larger and produce more electricity than those on

Figure 68 - Small windfarm in rural Sweden

land and can be kept from areas of natural beauty.

The energy transfer diagram would thus be:

Heat energy from the Sun -> E_p -> E_e -> electricity

11.5 Other schemes

WAVE ENERGY – Generators driven by up-and-down motion of waves.

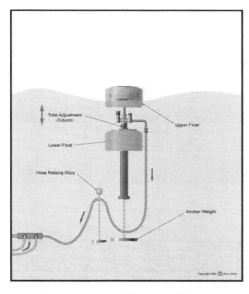
Upper float lowered in trough of wave

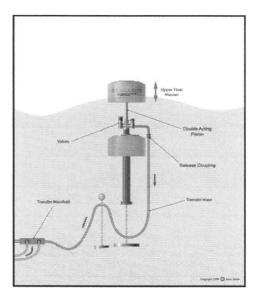
Upper float elevated on crest of wave

Figure 69 - Some wave power designs use the up and down motion to produce electricity

SOLAR ENERGY – Photovoltaic cells absorb radiation energy from sun. At A-level you are likely to learn about the only method of producing electricity that does not use a generator. It directly converts the rays from the sun to electricity in a process called photo electricity. This method is very inefficient and requires huge areas to gather the energy.

The energy transfer diagram is thus:

Electromagnetic Energy from the sun -> electricity

Figure 70 - Solar Farm

BIOFUELS – Fuels made from BIOMASS (plant and animal matter) and work in much the same way as those for non-recyclable fuels. The difference is that the fuels have been produced as a part of the natural carbon cycle and so are considered to be carbon neutral

GEOTHERMAL – Steam from turbines produced from thermal energy in earth. Water is pumped down into a hot region of the Earth's crust, where it gains heat energy and rises up through a separate pipe where it is used to spin a generator, thus producing electricity. The energy initially came from the formation of the Earth in the early solar system as it gathered together due to gravity. The Earth was initially very hot, and is cooling down, but the core and therefore the crust is still hot.

11.6 Renewable energy resources

The table below briefly details the advantages and disadvantages of the various renewable energy resources:

Energy source	Advantage	Disadvantage
Solar	• Free. • Renewable.	• Inefficient. • Expensive. • Variable sunshine.
Geothermal	• Renewable. • Free.	• Drilling expensive. • Geologically active areas only.
Tidal	• Renewable. • Free.	• Expensive. • Few areas suitable.
Wind	• Renewable. • Environment-friendly	• Noisy • Expensive • Hilly or wide open areas needed • Unsightly • Inefficient on land
Biofuels	• Varied sources. • Renewable. • Carbon Neutral	• Large land areas needed.
Wave	• Renewable. • Free.	• Little success. • Expensive. • Variable waves.
Hydroelectric	• Renewable. • Free.	• Expensive. • Few areas suitable. • Environmental damage.

Figure 71 - geothermal power plant in Iceland, where the Earth's crust is thin

Chapter 12. Energy, work and power

12.1 Work

Energy is the ability to make changes to something. In order for any changes to occur there is a need for some kind of force to act. Work is done when a **force** makes something **move**. More **work** is done when the force is **larger** and the distance moved is **further**.

Because:

> 1J of work done when a force of 1N moves an object 1m in the direction of the force.

$$Work \propto Force \text{ and } Work \propto distance\ moved$$

	Quantity	Name	Units
$W = F * d$	W	Work	Joules (J)
	F	Force	Newtons (N)
	d or s	Distance in the direction of the Force	m/s²

e.g. if a 4N force moves an object 3m, work done = 12J. The unit for energy, or work done, is the Joule, named after the Salford born physicist and brewer, James Prescott Joule, who studied the nature of heat and its relationship to mechanical work. He died in Sale, Cheshire, aged 70, the 11th of October 1889.

Figure 72- James Prescott Joule

12.2 Questions

1) Calculate the work done when a force of 254N moves an object 300m.
2) Calculate the force needed to produce 70J of work when moving an object 2m.
3) Calculate the distance moved when a force of 4N produces 8060J of work.
4) Express the following amounts of energy in Joules:

(a) 10kJ (b) 35mJ (c) 0.5mJ (d) 0.2kJ.

12.3 Power

Power is the rate at which work is performed. For instance, if I moved an object weighing 10N up 1m in 1s, I would be performing more work each second than if I moved it up the same distance in 10s. Power is measured in Joules per second, also known as Watts.

$$Power = \frac{Work\ Done}{Time\ Taken} \text{ or } \frac{Energy\ Transferred}{Time\ Taken}$$

$$P = \frac{E}{t}$$

Quantity	Name	Units
P	Power	Watts (J/s) This could be kW or MW or even GW
E	Energy	Joules (J)
t	Time	Seconds (s)

1W means that **work** is being done at a **rate** of 1J/s.

It is possible to calculate power when an object lifts a force through a vertical distance in a certain time using this equation. The vertical force is caused by the weight of the object due to gravity.

Worked example
A crane lifts a load of 3000N through a height of 5m 10 seconds. What is the power of the crane?

Work done = force x distance

= 3000N x 5m

= 15000J

Power = work done ÷ time taken

= 15000J ÷ 10s

= 1500W = 1.5kW

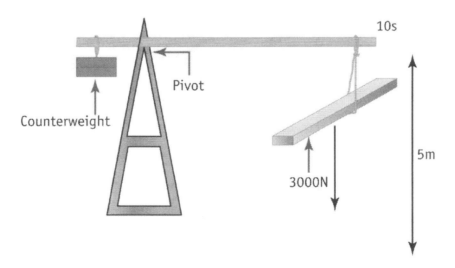

Figure 73 - Calculate the power

12.4 Electric power

Electrical energy also performs work. It is also measured in Watts. Electric items marked with electric power that they **transform** to a different type of energy.

Worked example
An electric kettle rated 2kW. How many joules of energy are transferred in 10 seconds?

Power = 2kW = 2000W = 2000J/s.

Rearranging the power equation to make the Energy the subject we have:

$$E = P * t = 2000 * 10 = 20000J, or\ 20kJ$$

12.5 Questions

1) A crane lifts a 600kg mass through a height of 12m in 18s. Calculate:
   a. The weight of the mass.
   b. The work done. (J)
   c. The power output. (Convert to kW).
2) A girl does 1000J of work in 5s. Calculate her power. Convert to kW.
3) A lamp is rated 100W. How many Joules of electrical energy does it convert into heat and light (a) each second (b) in 63s?
4) A motor has a useful power output of 3kW.
   a. Calculate to watts.
   b. Calculate useful work done in 1s.
   c. Calculate useful work done in 150s.
   d. Calculate efficiency if the power input to the motor is 4kW

Chapter 13. Momentum

When a mass is moving it possesses a quantity called **momentum**. This quantity increases as the mass increases and as the velocity increases. It is proportional to both and so is calculated using:

$$p = mv$$

Quantity	Name	Units
p	Momentum	Kgm/s
m	Mass	Kg
v	Velocity	m/s

E.g. is a car of mass 1200kg is travelling at 30m/s its momentum is 1200*30=36000Nm. Since the equation multiplies a vector (velocity) with a scalar (mass), the result is a vector. As such momentum is a vector with both size and magnitude. In physics there are a number of quantities that much be conserved, e.g. the conservation of energy. Unless an external force acts on a mass, the momentum must also be conserved. In fact, this connection leads to the original definition of Force, as defined by Sir Isaac Newton nearly 400 years ago when he said the force acting on an object was proportional to the rate of change of momentum:

$$F = \frac{mu - mv}{t} = \frac{m(u - v)}{t}$$

$$\text{since } a = \frac{u - v}{t}$$

$$F = ma$$

Quantity	Name	Units
F	Force	N
m	Mass	Kg
v	Final Velocity	m/s
u	Initial velocity	m/s
t	Time	S
a	Acceleration	m/s²

13.1 Collisions, Impulse

The quantity **impulse** is defined to the F*t, where F is the force acting on an object for a time, t. It would make sense that the longer a force is acting and the stronger the force, the larger the change in momentum, as defined above:

$$Ft = \Delta p = mu - mv$$

Quantity	Name	Units
F	Force	N
p	momentum	Kgm/s
v	Final Velocity	m/s
u	Initial velocity	m/s
t	Time	S

When considering this in a very practical situation, a car with a mass of 1200kg collides with a wall. The car stops in 0.05s. Working out the force on the car is then calculated by:

$$F = \frac{\Delta p}{t} = \frac{m(u-v)}{t} = \frac{1200(0-30)}{0.05} = -720000N = -720kN$$

Since Newton's third law applied to the passengers that are strapped in to their safety belts, the belts will exert that same force on the passenger.

Car designs therefore add crumple zones at the front and back of the car to allow the stopping time to be increased by causing the time the change of momentum changes to lengthen. In the same crash as above, with a crumple zone the time taken to stop is increased to 0.3s

$$F = \frac{1200(0-30)}{0.3} = 120000N = 120kN$$

By providing a crumple zone the forces experienced by the passenger from the safety belt is thus reduced by a factor of 6 and the injuries are likely to be much less sever.

Figure 74 - The front of the car crumples in, reducing the force on the passengers

13.2 Conservation of Momentum

We started off defining the idea of changing of momentum as only being as a result of a force. What happens then when objects hit other objects? Conservation is conserved in "elastic" collisions. Elastic collisions are when the kinetic energy is conserved in a collision, i.e. all the energy stays as kinetic energy and is not transformed to other types, e.g. sound or heat.

So, in an elastic collision the sum of the moments at the start must be the same as that at the end. In the following example, a ball of mass 2kg moving at 3m/s collides with a stationary ball with mass 4kg. What will their velocity be at the end if the two balls end up joined, and moving, together?

$$Momentum\ before = 2*6 + 4*0 = 12 kgm/s$$

After the collision, the new mass is 6kg:

$$momentum\ before = momentum\ after, so\ v = \frac{p}{m} = \frac{12}{6} = 2m/s$$

13.3 Questions

1. What is the momentum in kgm/s of a 10 kg trolley traveling at 5 m/s?

2. What is the momentum of a bullet of mass 75g travelling at 300 m/s?

3. What is the momentum of a human (mass 76 kg) fired out of a canon at 120 m/s?

4. Mike has a mass of 64Kg, is running at 5 m/s and jumps onto a 5 kg skateboard moving in the same direction at 4.2 m/s. What is their speed after Mike balances on the skateboard?

5. A truck of mass 750 kg moving at 7 m/s to the right collides with another truck of mass 1500 kg moving at 5 m/s to the left. What is their common velocity after the collision if they stick together?

6. Ben is standing still and fires a rifle. The bullet has a mass of 50g and is travelling at 330 m/s. If Ben has a mass of 70 kg, with what velocity does he move backwards.

7. A comet of mass 1×10^6 kg hits the earth at a speed of 1×10^5 m/s. At what speed does the earth move backwards after the collision (mass of earth = 7×10^{24} kg)? (You may assume that the Earth is stood still in order to solve this question)

Chapter 14. Astronomy

We shall explore this interesting topic from the perspective of how things evolved from the very beginning of time, gradually working back to our Solar System, the home of our home planet, Earth. Distances that are on astronomical scales are measure in **light years** – this is the distance light travels in one year in a vacuum and is approximately 1×10^{16}m – it takes light 8 minutes to travel from the Sun to the Earth, and about 1s from Earth to our Moon.

14.1 The Big Bang and Age of the Universe

Sometime about 13.78 billion years ago all of what we know as the universe, space, time and radiation, was created in a hot glowing ball of radiation and matter. Several phases have been identified by scientists;

Phase 1: <u>Very Early Universe</u>: It is so hot that only forces exist, not even particles
Phase 2: <u>The Dark Ages</u>: Once it has cooled down enough sees particles such as proton and electrons forming, and then becoming the lightest elements, especially hydrogen. This phase is known as the **dark ages**, since there were no stars and the universe had cooled too much for light to be created.

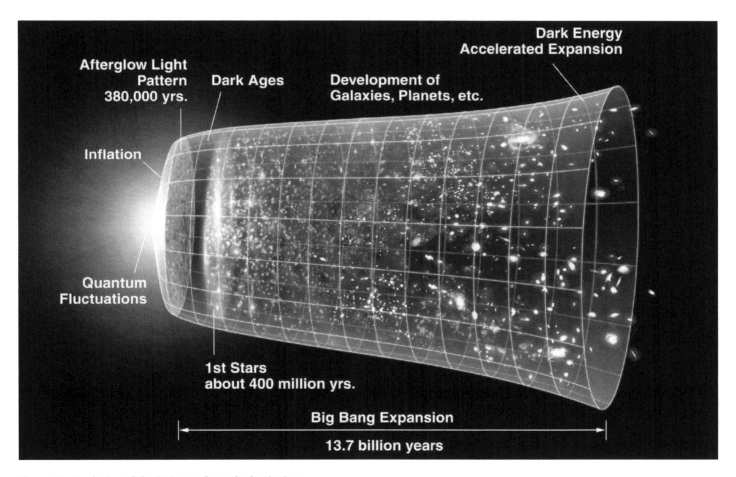

Figure 75 - Evolution of the Universe from the beginning.

Formation of stars and galaxies

Phase 3: <u>Structure Formation</u>: The Dark Ages ends up forming large quantities of hydrogen, helium and lithium clouds. Gases are pulled into areas where the gases already have higher densities, gradually forming hot gases that are so hot that they lose their bonds with each other and even the electrons orbiting the nucleus. This is called a **protostar** and has started to emit light. The gravitational pull on surrounding gases increases its mass and density and therefore the temperature. Once it has reached a high enough temperature it starts to **fuse** heavy hydrogen isotopes into deuterium and collapses further, becoming a star.

Figure 76 - A young star with its gas disc still around it

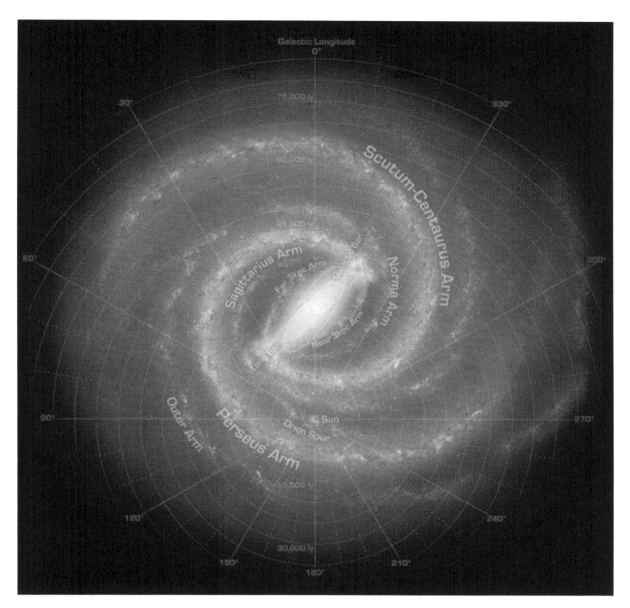

Once large numbers of stars form, they are pulled together into larger collection of stars, called **galaxies**. These can vary in size from thousands of stars, to hundreds of billions of stars. Our own galaxy is called the Milky Way and consists of about 200 billion stars (2×10^{11}). Scientists have discovered that there are about the same number of galaxies in the universe as there are stars in our galaxy.

The story continues as galaxies form into larger structures, called clusters. Our own galaxy is part of such a cluster, imaginatively called the **local group** and consists of 54 large and small galaxies. This structure is then part of a much large structure called a super cluster, which consists of about 100 clusters of galaxies. These super clusters are then organised into larger structures again called filaments – these look like long tendrils and, when enhanced through computers, they are very reminiscent of the way the neurons in the brain are connected (seeFigure 79 - Filaments that form the universe. Each dot of light is a super cluster. But these filaments are enormous, some are billions of light years across.

14.2 The Universe

As we zoom out of from the Earth, into the larger universe, we can see just how small, and insignificant our planet is. It is estimated that there are 6×10^{22} planets in the universe. Ours is, to date, unique in that it is the only one that has been found to have intelligent life on it. The distances involved to travel between stars that might have suitable planets on are so long that the time it would take to travel, or even to observe a signal from, makes it unlikely that we will ever interact with other worlds.

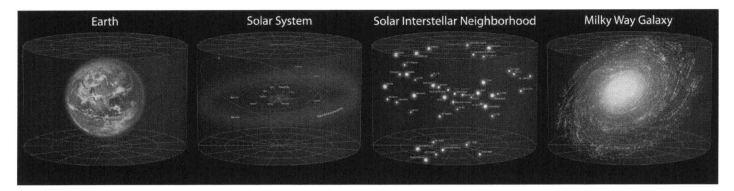

Figure 77- Zooming out from Earth to the Solar System, our local stars and our place in the Milky Way

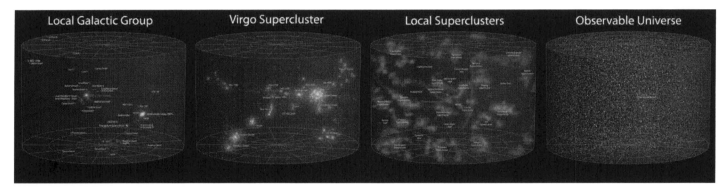

Figure 78 - Continuing to zoom out from the Milky Way to our local group, the super cluster we are part of and then further out again.

Recent work done by scientists looking at how the matter of the universe is distributed has found that the universe is riddled with filaments. Each light spot in the image is not just a galaxy, or a cluster, but super clusters. They appear to gather in strands and are huge – billions of light years in the most extreme cases.

14.3 The life of a star

Stars are initially formed as gases clump together in the centre of a spinning disc of gas and form a protostar of hot gas. When this gas is hot enough because of the pull of gravity, its nuclear fusion process is initiated in a violent explosion of energy. This pushes the gas closest to the star away and leaves the space immediately around it free of everything except heavy rocks.

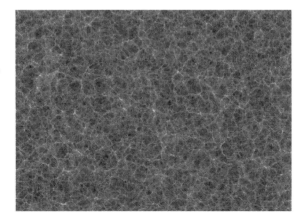

The star then uses its hydrogen fuel and depending on how massive it was, its later life may go one of three ways depending on how heavy it was.

Figure 79 - Filaments that form the universe. Each dot of light is a super cluster.

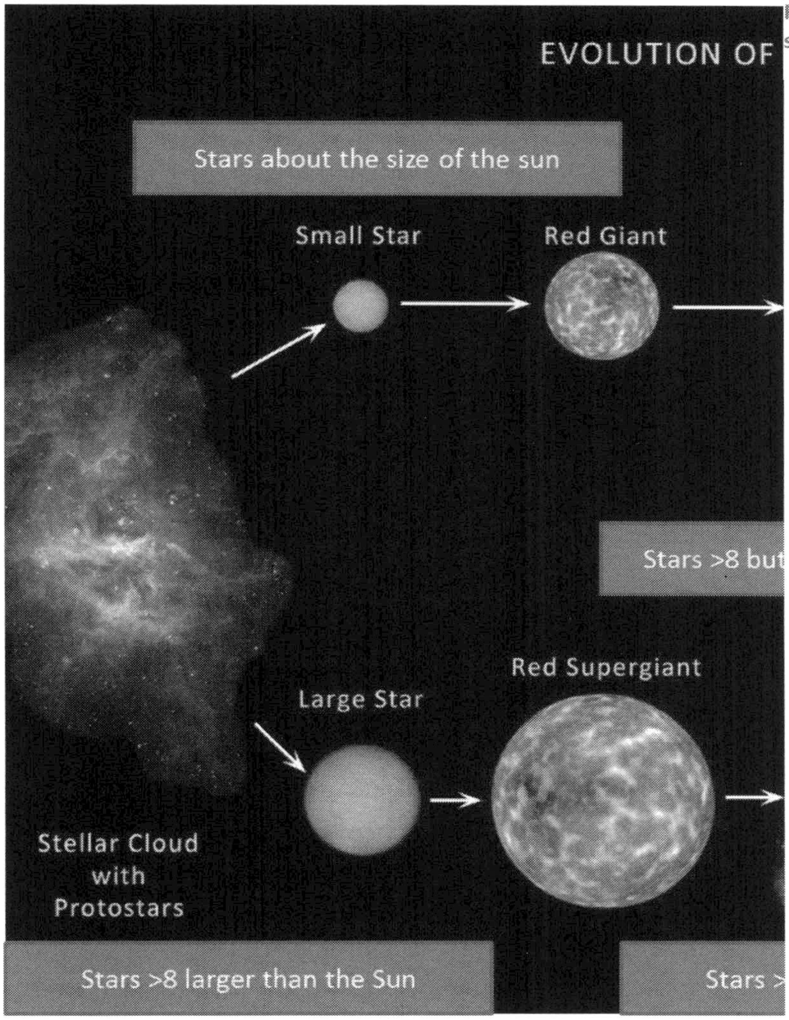

Figure 80 - Stellar life cycles for most stars

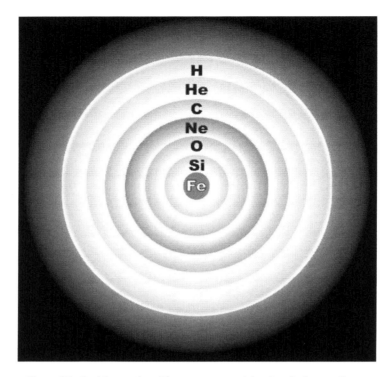

Figure 81 - Red Supergiant Element composition just before collapse

If it was about the same size as the Sun it will swell, and turn red, becoming a red giant as it starts to fuse the hydrogen and helium to heavier objects. The balance of the pressure from this fusion and that of gravity is kept stable until the star runs out of hydrogen at its core. It then, initially, collapses as gravity wins the battle of the forces. However, this collapse increases the temperature of the core 10 times and the outer part of the star expands to become a cooler, red giant. The inner part of the star is now fusing helium to heavier elements such as oxygen and carbon. Ultimately the cooler outer red giant shell, which is an unstable type of star, will blow off all its outer layers of gases, forming a planetary nebula (more correctly known as a stellar remnant nebula), leaving the central white dwarf exposed. As this star cools down it ultimately stops producing heat or light and becomes a black dwarf.

If, however, the start is much bigger than the sun, it has a more explosive end. Instead of becoming a red giant as it starts running out of hydrogen, it becomes a red supergiant, and instead of just the two layers that a red giant consists of, a supergiant has several layers, with the innermost layer consisting of iron, gradually becoming light elements until the outermost layer of hydrogen. When this collapses it undergoes a catastrophic explosive reaction to the collapse, called a supernova. This blows off huge amounts of gases into the interstellar space (which, because of the extreme heat, fuse to form very heavy elements, such as lead and uranium), and the remnant collapses to form either a neutron star, or a black hole, if the star was big enough. It is thought that most galaxies have a supermassive black hole in their centres and further black holes within the galaxy itself. A black hole has a gravitational field strength that is so strong nothing can escape from it, not even light.

14.4 Knowing these things

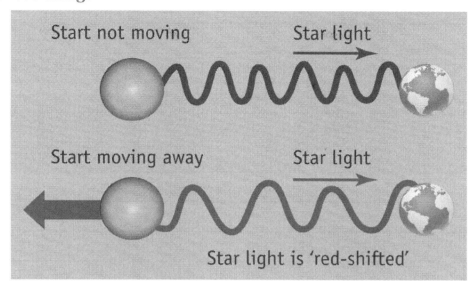

Figure 82 - How the doppler effect stretches light waves

Figure 83 - Edwin Hubble

Science often builds upon findings over long periods of time and understanding the development of the universe is one of the sciences that demand most understanding of physics. Edwin Hubble, seen in the picture, noticed that galaxies that were very distant were redder than expected. He also noticed that the further away a galaxy was, the redder it was. He used this together with a strong understanding the electromagnetic spectrum to realise that the colour had shifted to the red, called **red shift** as a result of racing away from the observer – this is an effect called the **Doppler effect** and can be heard when an ambulance is moving toward you and then leaving you; the pitch of the sound is higher on the way towards you than it is on the way away. The redder a galaxy is, the faster it is moving away. His conclusions were stunning to the scientific world:

1) Distant galaxies were moving away from us
2) The greater the distance away from us, the faster the movement away from us

The stunning conclusion, drawn from the fact that the Earth has no special significance in the universe, is that all objects are moving apart – **the whole universe is expanding**. The implication of an expanding universe was even more shocking. If the universe is expanding now, it means that it was smaller in the past. Looking at the movement of the stars away it was possible to work out when all stars were in the same place and this is thought to be about 13.78 billion years, a point in time called the **Big Bang**, a time of intense heat and release of intense gamma radiation. Of course, since we are looking at light start started off it journey billions of years ago, we are looking back in time also when we observe these, redder, galaxies.

Another common feature of good scientific discovery is that of taking a hypothesis and trying to either validate it, or disprove it by exploring it in more detail. If, it was argued, the universe was hot to start off, it had to radiate lots of high energy gamma rays. These would have been stretched from that point 13.78 billion years ago, and so should now be have become microwaves instead. So scientists started to look for this **cosmic background radiation** and found it (shown below, in Figure 84 - Cosmic background radiation in the microwave spectrum)! The Big Bang Theory had supporting evidence from other, reliable, observations.

Figure 84 - Cosmic background radiation in the microwave spectrum

More recent, validated, observations on distant galaxies indicates that there is an acceleration (see above) of the speed of the expansion of the universe and so it would seem that the universe will end in a huge, endless, yawn.

14.5 Questions:

1) What units do we measure distances in astronomy? Why do you think it is useful to use this unit instead of metres, or kilometres?
2) What is redshift? What is it about redshift that shows that the universe is expanding?
3) List the following in order of size
    a. Super Cluster
    b. Planet
    c. Star
    d. Cluster
    e. Planetary Nebula
    f. Filament
    g. Red Giant
    h. Galaxy
    i. Red supergiant
4) What would happen to a star that is:
    a. 100 times larger than our sun at the end of its life?
    b. 2 times larger than our sun at the end of its life?
    c. 15 times larger than our sun at the end of its life?
5) What is fused inside stars like our sun?
6) What two pieces of evidence support the theory of the Big Bang?

14.6 Satellites

When the inward force from the weight of the satellite is equal to the reaction force from the satellite that is pulling it outwards, it ends up orbiting a planet or star. The satellite, whether **natural**, such as a moon around a planet, or a planet around a star, has a velocity which is straight ahead, but the force of the gravitational body pulls it inwards. From the circular motion section, above, we know that the inward force is called a centripetal force. The outward force is caused by the object changing direction; according to Newton's Third Law (an action has an equal, an opposite, reaction).

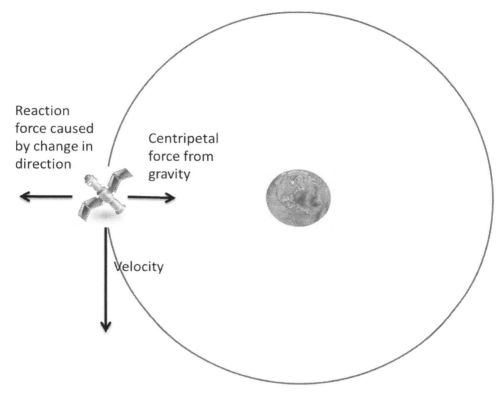

Figure 85 - Centripetal force caused by gravity causing a satellite to orbit a planet

The time taken to orbit (known as its **Period**) the gravitational body is determined by how far out it is from that body. Moving further out will cause that period to lengthen. If a satellite is designed to stay over one part of the Earth, it will have to have the same period as the Earth's rotation, i.e. 24 hours. This kind of satellite is called **geostationary**. Another type of orbit is one that goes across the poles and is by design lower and is called a **polar orbit**. Satellites that are sent up to perform specific jobs are called **artificial satellites**.

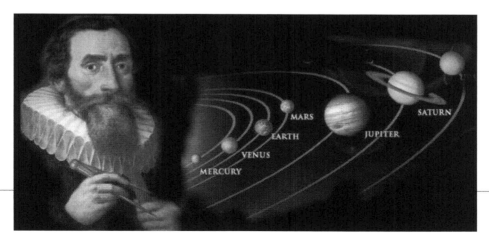

Figure 86 - Johannes Kepler 1571-1630 is credited with the physics of satellite motion

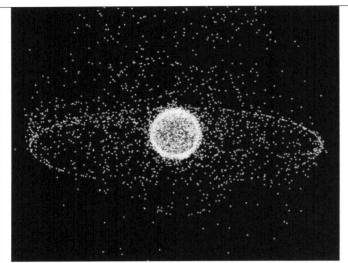
Figure 87 - The space around Earth is filled with satellites. The ring is where all geostationary satellites must orbit

Examples of **artificial** satellites:
- Telecommunications
- Weather predictions
- Military purposes such as spy satellites
- Satellite navigation such as GPS
- Scientific Research
- Imaging satellites, taking pictures of earth, e.g. for Google Earth™

Examples of **natural** satellites
- The Sun around the centre of the Galaxy
- Venus, Earth and Jupiter around the Sun
- Comets around the Sun
- The Moon around the Earth
- Dwarf planets such as Pluto, Charon and Eris

Figure 88 - A comet. The tail always points away from the sun

Satellites must be much lighter than the object that it is orbiting since the larger object is changing the direction of the smaller, and lighter, object. The force exerted by the smaller object is not enough to fully change the direction of the heavier one.

Orbit speeds of satellites

Artificial satellites vary in how rapidly the move around a planet, from 3000m/s for telecommunications satellites to 8000m/s for those orbiting lower, in polar orbits. Calculating the orbital speed uses the idea of the distance moved around a circle, in terms of multiples of 2π, the multiple used to calculate the circumference of a circle based on its radius. The time it takes for a satellite is called its period (T), as outlined above. The speed of the satellite is therefore:

	Quantity	Name	Units
$$orbital\ speed = \frac{2\pi r}{T}$$	Orbital peed		m/s
	r	raius from the centre of the planet	m
	T	Time taken to orbit the planet (or star)	s

Worked example

Calculate the speed of a satellite that is orbiting 200km above the Earth's surface and completes one orbit in 1h and 24 minutes. The radius of the Earth is 6400km.

First we need the period in seconds: 1 hour and 24 minutes is 60x60 + 24*60=5040s

The radius of the equation above is to the centre of the Earth, so the radius is 6400+200km, or 6600000m.

$$Orbital\ speed = \frac{2\pi x 6600000}{5040} = 8200 m/s$$

14.7 The Solar System

The solar system contains a considerable host of objects that orbit the sun as natural satellites. Best known are the planets. The list includes; The Sun, the planets (of which there are 8), dwarf planets, asteroids and comets. There are also stranger objects that are very difficult to observe, such as centaurs and Neptunian Trojans.

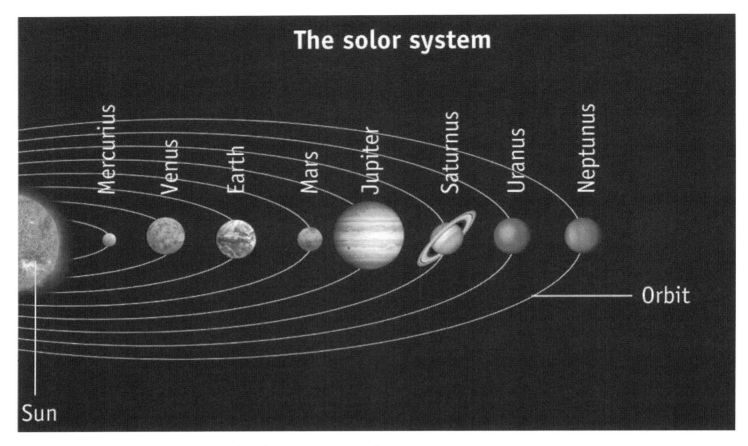

Figure 89 - The locations of the 8 planets in the solar system - not to scale

14.8 Planets

There are 8 planets orbiting the sun, four rocky ones found closest in towards the centre. When the solar system was initially formed, the gases in the central area were blown away by the initial formation of the sun. Only rocks were left. These gradually merged together to form these planets. We will explore them from the centre and outwards. When considering the distance from the sun the concept of an **astronomical unit (AU)** is used. Earth is one AU from the sun. Planets to not produce their own light, they are **non-luminous**, but can be seen because they reflect the Sun. As seen above, under the section for weight, each planet has different gravitational field strength (Table 4 - gravitational field strengths for the different planets in the solar system). On Earth we feel this as our normal weight, but other planets would make us feel heavier, or lighter. The lightest weight would be on Mercury, whereas we would be nearly 3 times heavier on Jupiter.

Mercury (0.4AU)
Mercury is very hot with an average temperature of +350°C and therefore devoid of life. The north pole of this planet has a very deep crater and recent research has shown that this has frozen water in it. The day is very long and so, because there is hardly any atmosphere, the temperature on the "back" of the planet can drop very low. Research has shown that because of this, there is some atmosphere on the back side.

Venus (0.7AU)
Venus is very similar to Earth in size, and has a very dense atmosphere made of dense, hot (>400°C), acidic, gases. Because of the hostile environment, the few probes that have been sent to Venus have been destroyed within seconds of entering its atmosphere. The ground is riddled with high volcanoes and lava flows. Venus is easy to see in the night as it reflects the light from the sun.

Earth (1AU)
Earth is the only planet with considerable amounts of liquid water and as such has intelligent life. Its moon is much larger than other planets and it is thought that it is as a result of a proto-planet colliding with proto-earth and knocking off a large section of the new planet about 4 billion years ago. Earth has one natural satellite, the Moon. It takes the Moon 29.5 days to orbit the Earth.

Mars (1.5AU)

Mars is considerably smaller than Earth and is being carefully examined at present to see if it might have had life. It is certain that it had flowing water in the distant past but that this evaporated off due to its low gravity. Mars has two natural satellites, Phobos and Deimos, both which are too light to have formed proper spherical shapes.

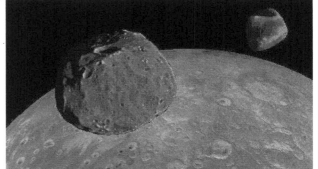

Asteroid Belt and Dwarf Planets

Once we leave the rocky planets behind we enter a region that is full of small rocks and some of these are large enough to be classed as dwarf planets. This region is called the asteroid belt and two of its largest asteroids are known as Ceres and Vesta. Asteroids are small, rocky, planets – often called planetoids. The image to the right shows these asteroids as being densely packed, however, the distances between each individual asteroid is typically very large, allowing space probes to pass between them to expore the outer planets, and beyond.

When the sun first formed in an explosive burst, the gases from near the centre were blown outward and formed the outer four gas giant planets.

Jupiter (5.0AU)

Jupiter is the largest planet of the solar system and still much smaller than the Sun. It is formed mainly from Hydrogen and Helium gas, although its core is likely to be twice the size of the Earth and made of iron. It has a very powerful magnetic field around it and is home of over 63 moons, 3 of which can be seen with a weak telescope. The size of Jupiter has made it a natural ally of Earth as any stray objects from outside the solar system is likely to be attracted to it before arriving at Earth, possibly causing havoc.

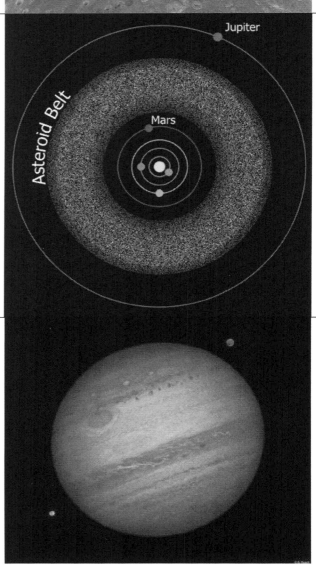

Saturn (9.5AU)

Saturn is smaller than Jupiter but is well known for its amazing rings. These rings are of unknown origin but might have been formed by the gradual release of ice from one of its closest moons, or from a moon that was destroyed by its gravity. It is mainly formed out of gases and, if it has a core, it is lightly to be very small. The gravitational field strength is similar to that on Earth. It has 62 known moons, most of which are very small.

Uranus (19AU)

Uranus would seem to be very unremarkable. Discovered in 1781 with a rudimentary telescope by Herschel, it initially was only just a rather featureless heavenly body. The best picture of it, from Voyager 2 in 1986, shows it to be a simple shape. However, it does have a magnetic field and a very feint ring system and 27 moons. Uranus is unusual because its rotation is tilted against the rest of the planets. This is possibly because it was hit by another planet early on in the formation of the solar system which caused it to change its rotation.

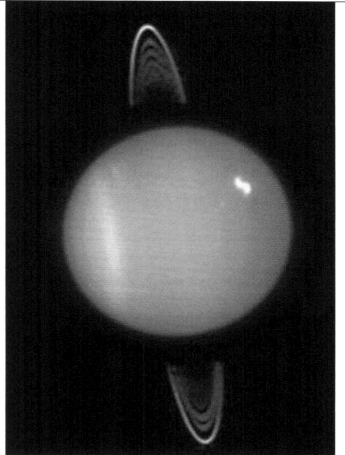

Neptune (30AU)

Neptune is the farthest planet from the sun and is made up of hydrogen and helium, with a small core. It is blue and the picture, taken by voyager 2 in 1989, shows it to have weather systems. It also has a very feint system of rings around it, as well as 13 known moons.

14.9 Orbiting times

As a rule, objects further away from the sun take longer to orbit the sun. Johannes Kepler (1571-1630), a German scientist, is credited with understanding why this was, and put together three laws that described how the motion of planets could be understood. These are beyond the scope of this book, and have since been superseded by Einstein's theory of general relativity. They were, however, instrumental in helping understand that the solar system is a result of purely mechanical principles rather than a divinely controlled structure.

Another rule that can be followed is that the further away from the sun, the colder the planet is, with the one exception of Venus. Venus is warmer than Mercury because of the thick layers of clouds keeping the heat in.

Table 5 - Planetary data

	Mercury	Venus	Earth	Mars	Jupiter	Saturn	Uranus	Neptune
diameter (Earth=1)	0.38	0.949	1	0.53	11.209	9.44	4.007	3.883
diameter (km)	4,878	12,104	12,756	6,787	142,800	120,000	51,100	49,500
mass (Earth=1)	0.06	0.815	1	0.11	318	95	15	17
mean distance from Sun(AU)	0.39	0.72	1	1.52	5.20	9.54	19.18	30.06
orbital period (Earth years)	0.24	0.62	1	1.88	11.86	29.46	84.01	164.8
mean temperature at surface (°C)	-180 to 430	465	-89 to 58	-82 to 0	-150	-170	-200	-210
number of moons	0	0	1	2	63	62	27	13
rings?	no	no	no	no	yes	yes	yes	yes

In between the gas giants

The distances between the gas giants are enormous and houses numerous heavenly bodies that are very specific to that region, including Centaurs (small solar bodies with a strange orbit around the sun), Neptune Trojans (asteroids that move around with Neptune at about the same distance) and detached objects from the Solar system.

The solar system beyond Neptune – The Kuiper Belt and the Oort cloud

There are numerous trans-Neptunian objects (TNOs) – objects further out than Neptune - that we are aware of, including the dwarf planets show above, and beyond this is another belt of small rocks, left over from the formation of the solar system, called the Kuiper belt, and then, further out a hypothesised cloud, where some of the comets originate, called the Oort cloud.

The Kuiper belt is huge, but otherwise similar to that of the asteroid belt, containing vast numbers of icy, rocky or metallic objects. It appears to stop very suddenly at about 50AU, for no apparent reason. The best-known Kuiper belt object is Pluto, and its moon, Charon. Beyond this belt, and surrounding the solar system in all directions is the Oort cloud. The Oorth cloud has never been observed, but it is thought that very eccentric comets (see below) originate from this very distant reach of our solar system. It is, however about 50,000 AU from the Sun, and thus about a quarter of the way to our nearest neighbouring star, Proxima Centauri.

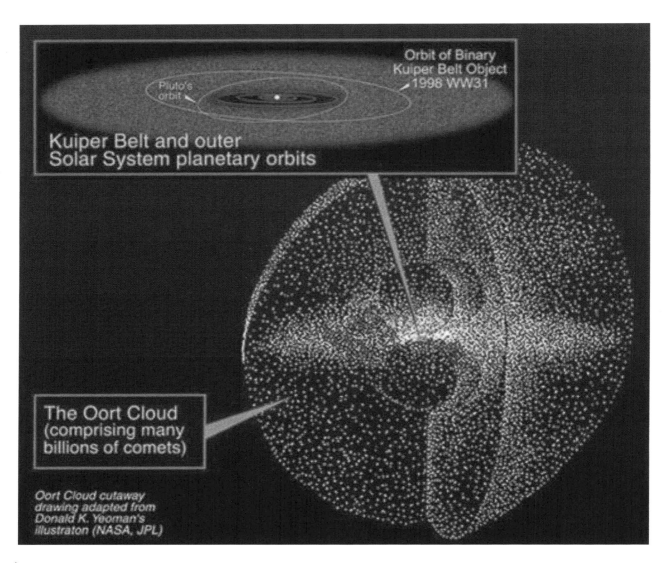

Comets

Comets are icy solar system objects that become visible when they get close to the sun. They usually have very eccentric orbits and return with regular time intervals, usually with 7 or more years between each visit. Eccentric orbits are those that are at strange angles to the plane that the planets orbit on, and one end of its orbit is very close to the Sun, the other end of its orbit can be well outside that of Neptune.

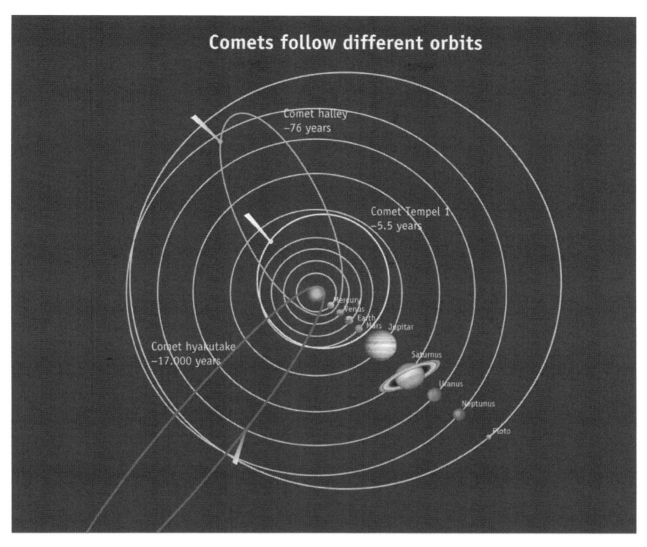

Figure 90 - Comets have eccentric orbits, come in from all kinds of angles and their tails ALWAYS point away from the Sun

14.10 Questions

1) How many planets are there in the Solar System? List them in order, from the Sun outwards
2) What is the definition of 1 astronomical unit?
3) Which are the four gas giants in the Solar System?
4) Which planet is closer to the Sun, Earth or Mercury?
5) A weather satellite has an orbital speed of $7780 ms^{-1}$, at an altitude of 300km from Earth (the radius of Earth is 6400km), what is its period in seconds?
6) Give three examples of artificial satellite uses and mention two different types of orbits.

Chapter 15. Astrophysics

Stars are classified many ways, specifically through observation of their colours and brightness's. A huge amount if information is hidden in these two areas, as we will see.

15.1 Colour

The colour of a star is directly related to its surface temperature. This is linked to something that can be measured in a lab using something that simulates a "black body" (an object that is a perfect absorber and emitter of energy). When measured a curve like the one below is found at different temperatures:

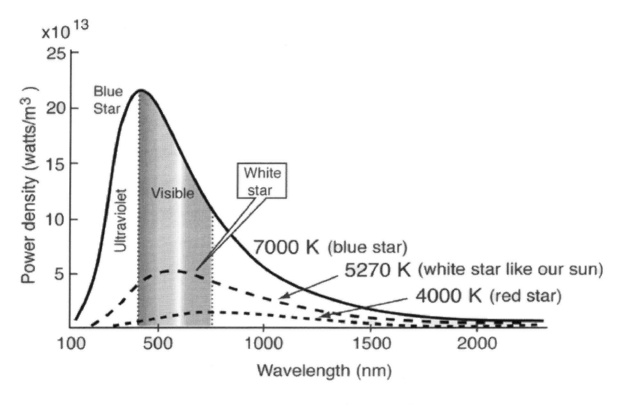

Figure 91 - Black Body radiation

When we transfer this to a table we can see how it converts to very specific stars (with names)

Table 6 - Star Standard Spectral Types

Type	Colour	Approximate Surface Temperature	Main Characteristics	Examples
O	Blue	> 30,000 K	Singly ionized helium lines either in emission or absorption. Strong ultraviolet continuum.	10 Lacertra
B	Blue	11,000 - 30,000	Neutral helium lines in absorption.	Rigel Spica

A	Blue	7,500 - 11,000	Hydrogen lines at maximum strength for A0 stars, decreasing thereafter.	Sirius Vega
F	Blue to White	6,000 - 7,500	Metallic lines become noticeable.	Canopus Procyon
G	White to Yellow	5,000 - 6,000	Solar-type spectra. Absorption lines of neutral metallic atoms and ions (e.g. once-ionized calcium) grow in strength.	Sun Capella
K	Orange to Red	3,500 - 5,000	Metallic lines dominate. Weak blue continuum.	Arcturus Aldebaran
M	Red	< 3,500	Molecular bands of titanium oxide noticeable.	Betelgeuse Antares

You will not need to memorise this, but you should have a general understanding of these, that red is cool and blue is hot. You can remember the OBAFGKM by the mnemonic of "Oh be a fine guy, kiss me".

15.2 Brightness

The brightness of the stars depends on a number of things, including how far away they are and the amount of energy they produce. The early Greek astronomers classified them based on how soon they were able to see them in the night sky. The first ones to show up in the evening were the 1st-order, and as the night closed in the fainter stars were visible, and the last ones were called 6th-order. These became magnitude 1 to 6 (the lowest brightness an eye can see). Odd, as the brightest stars then have the lowest value. Our own sun, being closest, therefore ends up with a negative magnitude of -27!

In 1856 Norman Pogson realised that although the magnitudes were subjective, he could see a pattern that the difference in brightness between each magnitude was 2.512, so the difference between 1 and 4 stars are 2.512^3, or 15.8 times brighter. In your syllabus this 2.512 has been rounded to 2.5.

However, as mentioned, the magnitude depends on how far away something is too. We use two terms to differentiate between the two.

- **Apparent magnitude,** the brightness of a star as seen from Earth. The Sun, when see from here is -27, but it not actually as bright as Sirius A with an apparent magnitude of -1.46.
- **Absolute magnitude**, the actual brightness of a star when corrected for the distance between Earth and that star. The way this is done, mathematically, is to place a star at a distance of 10 pc (parsec) or 32.68 light years. We then compare these stars and can compare their actual power output. Sirius A (part of a binary system with a much darker companion called Sirius B) is +1.42, the sun +4.83 and R136a1 (a star in the large megallanic cloud) is -12.5 (the most powerful star we have observed).

15.3 Hertzsprung-Russell diagram

In around 1910 Hertzsprung and Russell plotted a diagram where one of the axes was the temperature of a star (or the spectral type) and the other the absolute magnitude of the star. They noticed that there were essentially four major areas. The first is a diagonal line, called the main sequence, where stars exist during their main fuel fusing life, then, depending on their mass, they either form Giants of Supergiant. A fourth are is formed when giants die and form white dwarfs. You must be able to place these areas on the right part of the diagram. You should also note that the lifetime of a star depends on how hot it is in the first place.

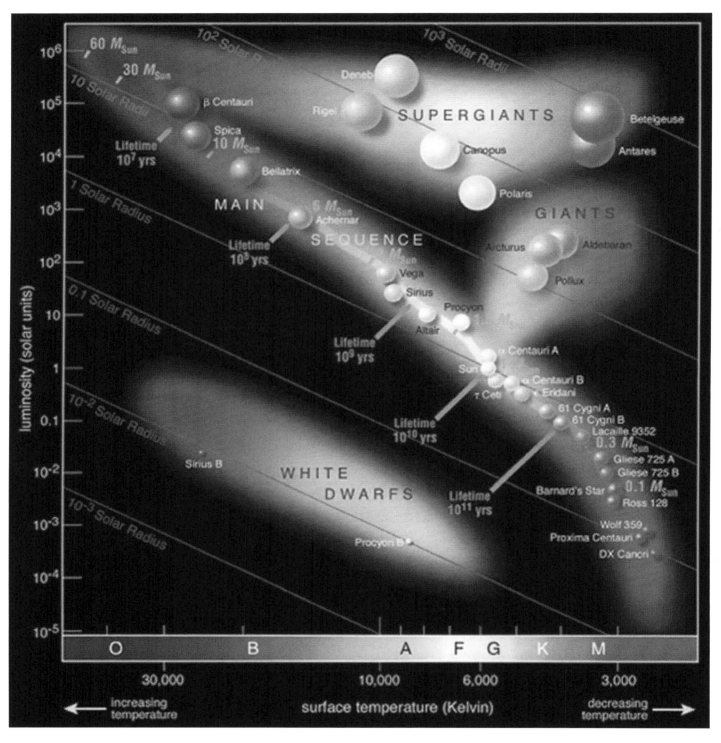

Figure 92 - Hertzsprung-Russell diagram

15.4 Blackbody radiation

All objects that have kinetic energy, i.e. that have a temperature above 0K (-273.15°C) emit electromagnetic radiation. A "Blackbody" is a theoretical object that emits and absorbs everything across all wavelengths of electromagnetic radiation. The resulting radiation can be predicted using theoretical and observed graphs at different temperatures. It turns out that stars are really close to be blackbodies. Using these curves at different temperatures, looking at the distribution and intensity of the different wavelengths, we can work out how warm a star is by matching it to the curve.

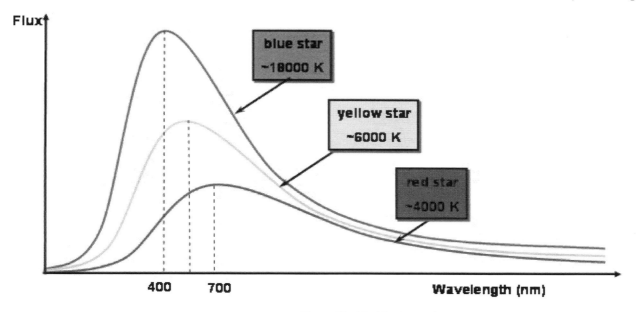

Figure 93 - blackbody graph

It is also useful in seeing how a body warms up over time. The Earth is a good example of an object that should be a blackbody. However, when an object absorbs more energy than it emits, its temperature increases. Greenhouse gases, such as CO_2, contribute to this and so blocks the heat from escaping, thus the temperature will increase, and so the Earth is not, presently, a blackbody.

Using these graphs, it was possible to work out the temperature of the Cosmic Microwave Background Radiation, and found it to be 2.725K, which is in line with an age of the universe of 13 billion years, another piece of evidence that supports the Big Bang theory.

Chapter 16. Cosmology

The study of the formation and development of the universe is called Cosmology and is a fairly new science, as we will see.

16.1 The Hubble Law and Doppler Shift

Early work in astronomy led to the assumption that the whole universe was our own galaxy, but in some of the earliest work of the world-famous Edwin Hubble, he realised that, what was initially a blurry feature, called a nebula, was in fact a new galaxy. We now know that as the Andromeda galaxy. He used new methods of determining how far away it was based on Cepheid variable stars (see later). In 1929 he was measuring the distances to galaxies much further away and realised that their light was stretched. This is a result of the Doppler effect (think of ambulances speeding towards you and then away – the pitch changes).

He noticed that nearly every galaxy was stretched into the redder part of the spectrum, indicating that they were moving away from us. The only way that could be true is if the universe was expanding. If the universe is expanding, then it follows that the universe was smaller in the past, and when that size of just about nothing, was when the universe was formed – the Big Bang. At the moment we calculate this to be 13.78 billion years ago.

The Doppler equation is:

$$\frac{\Delta \lambda}{\lambda} = \frac{v}{c}$$

Where λ is the wavelength of the light in the lab, and Δλ is the difference between the measured wavelength and λ, v is the speed of the object being observed and c is the speed of light. Hubble produced a diagram of the doppler shift against how far away the object is.

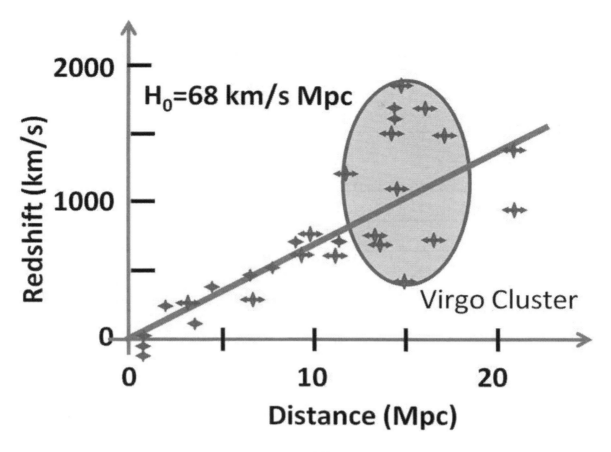

Figure 94 - Hubble's Law

The only way of explaining that an object twice as far away as another one is if the space between the galaxies themselves is expanding.

16.2 Evidence that demanded the Big Bang theory

The Steady State theory, developed by the cosmologist Fred Hoyle, proposed that the universe had always existed in a similar state to the way it appears nowadays. Although stars and galaxies are born, evolve and die, the average numbers of each has always remained the same. The average numbers of galaxies in our universe is the same no matter which direction we look from the Earth and the Steady State theory was suggesting that this symmetry extended to time itself, implying that our universe had no beginning and would have no end.

The Big Bang theory proposed that around 13.78 billion years ago our universe existed in an incredibly tiny space. It then began to expand very quickly – an event known as the Big Bang and we can still see this expansion taking place today, although gravity has brought matter together to create stars and galaxies in certain areas. Through the 1950s there was a fierce debate amongst cosmologists as to which of these two theories was the better explanation for the expanding universe. However, during the 1950s and 1960s there were two major discoveries which tipped the balance firmly in favour of the Big Bang theory. The Steady state theory still exists, but is very much a fringe part of cosmology, especially since it breaks the idea of creating energy (conservation of energy).

The two discoveries were: The Red Shift and the Cosmic Background Radiation.

The idea that **red shift** implies the universe is expanding, and therefore must have been smaller in the past, pretty much killed the idea of the steady state.

In support of this was the discovery of the **Cosmic Background radiation** (CMBR), accidentally measured by Arno Penzias and Robert Wilson in 1965. They noticed that whatever direction they pointed a radio telescope, they detected a "hiss" – and deduced that it was left-over radiation from the early universe, when electromagnetic radiation could first propagate. When measuring the strength of wavelength of this signal, it also provided and age of about 14 billion years ago that the radiation was formed. See Figure 84 - Cosmic background radiation in the microwave spectrum.

16.3 Dark Energy and dark matter

In 1998 Adam Reiss made observations of distant galaxies and noticed that, unlike the theory of a universe that either would continue to expand and then slow down, or was already slowing down under the influence of gravity, it was in fact increasing in the rate of expansion. One of the main explanations for this has been dubbed dark energy.

Observations of galaxy rotation indicated that there was much more matter in the galaxy than could be observed, as did a phenomenon called "gravitational lensing" when light is bent around a galaxy or cluster of galaxies, but it is curved too much for the amount of matter observed in the galaxies.

In both of these cases, scientists have yet to determine what is behind either. Many ideas have been proposed for both and scientists are working through each one.

Chapter 17. The kinetic model

All matter is made up of particles, which are arranged in specific ways:

1) They have attractive forces between them
2) They have specific types of structures
3) The move in certain ways

The key to understanding this model is to understand that each particle can be visualized as very small object – typically an atom, or a molecule, is represented as a small ball.

17.1 Properties of the states of matter

There are three states (of phases) of matter:

1) Solid
2) Liquid
3) Gas

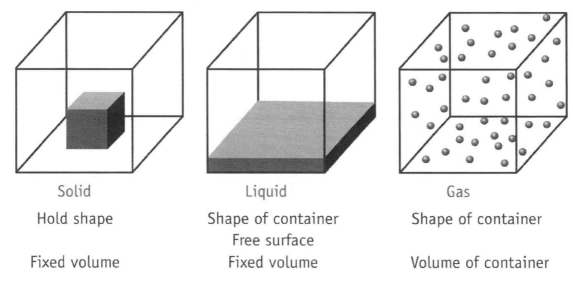

Figure 95 - The three phases of matter

Solid	Liquid	Gas
Have a fixed shape and volume Particles are close together There are strong forces of attraction between the particles They vibrate around a fixed position	Have a fixed volume (cannot be compress) Will take on the shape of its container (can flow) Particles are close together and have weak attractive forces. They move about each other	Has no fixed volume or shape (can flow into any shape) Fills any available space Particles are fast moving and spread out Very weak attractive forces

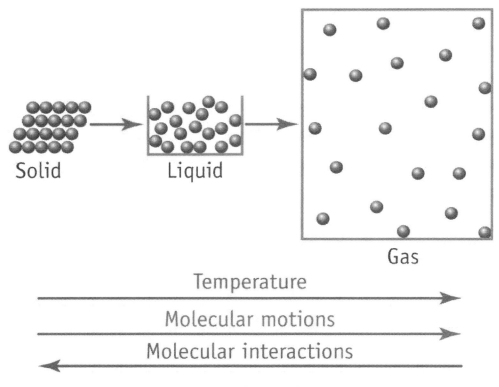

Figure 96 - Comparisons between the three phases

17.2 Evidence for moving particles

Most, if not all, scientists accept the concept of the kinetic molecular model. Any good scientist must, however, be willing to prove any idea is accurate through some form of observation. This is also true for the idea of moving particles. One of the main pieces of evidence for this comes from a simple experiment that shows a phenomenon called "Brownian Motion".

In Brownian motion, it is easy, through a microscope, to observe a random motion of smoke particles, as these bounce off water molecules in the air.

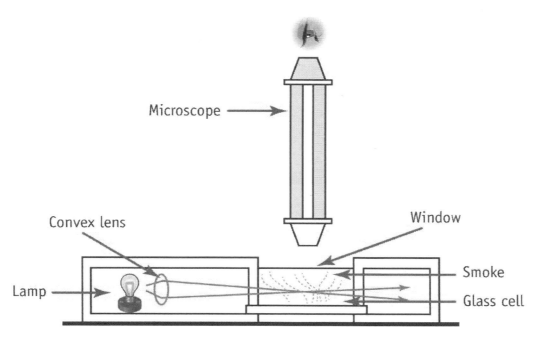

Figure 97 - exploring Brownian motion. A glass container is filled with smoke. The smoke particles can be seen bumping off other, invisible, air particles.

In this experiment, the smoke from a paper straw is blown into a glass cell and covered with a glass window. A lamp is lit to illuminate the small smoke enclosure. A travelling microscope is place above and the smoky area is observed. A random movement of particles is observed. The explanation for this is that each smoke particle collides with air and water molecules, and, as a result, changes direction.

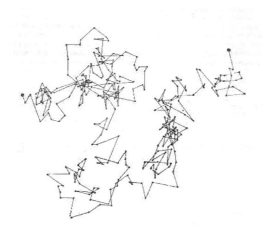

Figure 98 - A typical path of a smoke particle when observed in the apparatus shown above

17.3 Movement and energy of particles

All particles observe the usual laws of having kinetic energy (K.E.) as a result of movement, and gravitational potential energy (P.E.) as a result of their position. Gas molecules have the highest K.E. and also, usually, the highest P.E. The Internal Energy of a particle is the total of the K.E and P.E. As the temperature of a compound increases, so does the Internal Energy. This is because the particles start moving faster.

The Kelvin Temperature Scale

When working with Gay-Lussac and Charles' laws, it is important to use the absolute temperature scale, the Kelvin scale, as this reflects the actual kinetic energy in the gas. This scale starts at -273°C, which is called **absolute zero** and is 0K. This scale is used to measure very high and very low temperatures. It is not possible to go lower than 0K.

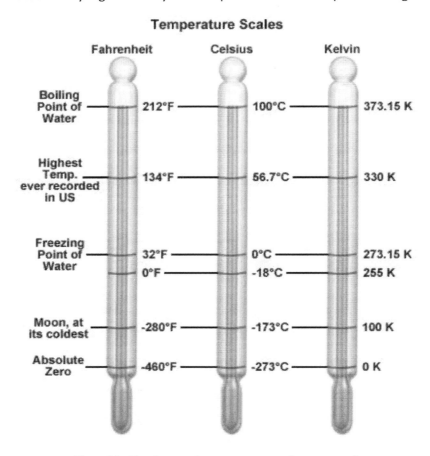

Figure 99 - The three main temperature scales compared

17.4 Boiling and Evaporation

Two methods of changing from a liquid phase to a gaseous phase:

- Evaporation – random energy levels in the liquid causes the top layer to become a gas

- Boiling – the heat in the liquid has reached the level where the particles are energetic enough to form gas bubbles that will rise.

Evaporation

The energy within matter is not perfectly distributed; some particles have more energy than their neighbours. If this happens at the surface of a liquid, the particles can escape the weak bonds of the liquid and become gaseous. This process is called evaporation. The rate at which evaporation can occur can be affected by changing:

1) The **temperature** of the liquid – this increases the internal energy and therefore the kinetic energy

2) The **surface** area – This causes more molecules to be close to the surface, thus increasing their abilities to turn gaseous

3) The existing **humidity** if the air - if there are already water molecules in the air, it makes it more difficult for other ones to join them.

4) **Wind** movement – if wind removes the higher energy water particles, which would otherwise block the release of newly evaporated particles, it becomes easier for the next layer of surface particles to become gaseous.

5) Decreasing the **pressure** upon its surface

Cooling effect of evaporation

Giraffes lick their shins when it is warm to help them cool down. In a similar way, if you dampen the back of your hand and wave it about, you will feel a cold sensation. This is because the process of evaporation carries away the higher energy particles and therefore leaves the lower energy ones behind. Lower energy particles are cooler. This type of energy is called Thermal Energy and is called the Latent Heat of Evaporation. It is used in nature in sweating and in refrigerators.

Sweating

When animals sweat, the fluid will end up evaporating, thus removing thermal energy from the creature.

Refrigerators

Fridges use a "heat pump" in reverse to move thermal energy from a colder environment out into a warmer environment. Since the temperature of a body is a measure of the average kinetic energy of its particles. During evaporation, the molecules which are more likely to "escape" from liquid and become part of the vapour are the ones which have higher than average kinetic energy. Therefore, if you cause the rate of evaporation of a liquid to increase, without supplying energy, the temperature of the remaining liquid will decrease.

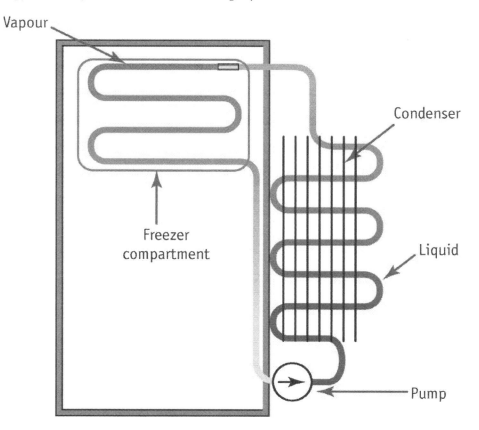

Figure 100 - A heat pump that removes heat from the cold inside of a fridge to the warmer outside

In the tubes around the freezer compartment, the pressure is decreased by the pump (there is a small section of the tube which is narrower than the rest). Rapid evaporation takes place here and latent heat of vaporisation is taken in.

In the tubes outside the refrigerator, the vapour is compressed and then it condenses. Latent heat is given out as it condenses.

1. As the liquid goes through the condenser, it cools, removing thermal energy
2. At the top there is a small valve that forces it evaporate into a gaseous form and it simultaneously drops on temperature
3. As the cooled vapour passes through the freezer, thermal energy from the freezer box is transferred to the vapour
4. The vapour is then passed through a pump, which compresses it, turning it back into a fluid and passing it through the condenser.

The full theory of this is covered in applied physics, Unit 5, of the AQA A-level syllabus.

Boiling

When enough energy has been absorbed by a group of neighbouring particles, they care able to separate from each other and become gaseous. This can happen within a fluid. Once this happens, bubbles form with the vapour within the fluid. These bubbles rise to the top of the surface and as they do so they expand. As they are released from the fluid the burst and the vapour is released into the surroundings.

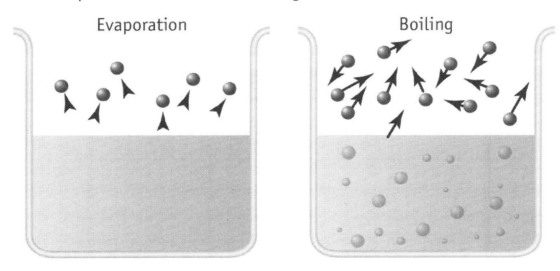

Figure 101 - How boiling and evaporation differ from each other

17.5 Questions

Explain:

1. Why, on a humid day, you feel hot and uncomfortable?
2. Why puddles dry out quickly
3. How the kinetic theory explains the cooling effect
4. What is the difference between boiling and evaporation

17.6 Thermal expansion

When matter is heated, they, typically, expand. This means that they increase their volume due to heating. All three states of matter expand (with a minor exception for water between 0C and 4C). This expansion is explained by the kinetic molecule theory:

1) As each particle gains more K.E., it moves faster (in solids, they vibrate faster)

2) As it moves faster, it takes more volume by itself (in solids, this vibration is larger)

3) All neighbouring particles also increase in volume for the same reason (in solids, each vibrating particle pushes all the other ones further away)

4) Therefore the whole compound increases in volume

The hotter the matter, the larger its volume will be as a result of heating. In solids, this expansion occurs in all directions, and can cause a solid to stretch a considerable amount (see below).

> *Invar, also known generically as FeNi36 (64FeNi in the US), is a nickel iron alloy notable for its uniquely low coefficient of thermal expansion. The name, Invar, comes from the word invariable, referring to its lack of expansion or contraction with temperature changes.*
>
> *It was invented in 1896 by Swiss scientist Charles Édouard Guillaume. He received the Nobel Prize in Physics in 1920 for this discovery, which enabled improvements in scientific instruments.*

17.7 Expansion in Solids

Using expansion

Bi-metallic strips

These devices, which are reminiscent of the old style metal spatulas, consist of two metals that expand different amount as a result of being heated. They are fixed to each other length-ways. As the two strips are heated, the one that expands is forced to bend as a result of being constrained by the other strip. The result is that it will bend inwards.

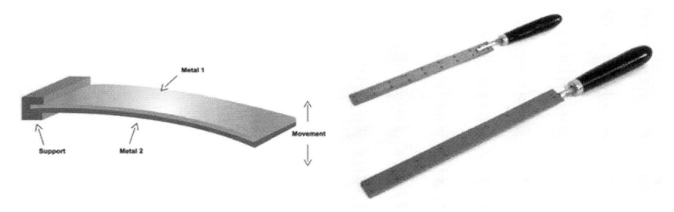

Figure 102 - How a bimetallic strip might bend and how they are shown in use in schools

These two strips can be made of iron and brass, where brass will expand more, thus causing it to bend around the iron strip.

Fire Alarm

A circuit with a bimetallic strip is created, where the strip will act as a switch. It is connected to one side of the circuit and, when heated, it will bend to connect to the other part, thus closing the circuit, and cause the bell to ring.

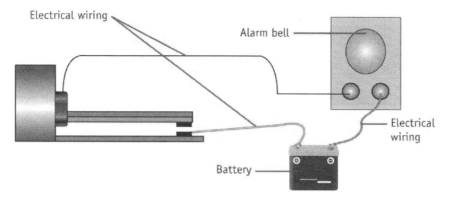

Figure 103 - Fire alarm using a bimetallic strip to detect when it is too hot

Electric thermostat

A similar circuit to a fire alarm with a bimetallic strip that will open when the temperature is too hot, thus turning off a heater. When the temperature is lower, the strip will straighten and re-establish contact, turning on the heater. This will help establish a stable temperature, in a room, for instance. A screw has been added that pushes the half of the switch that is not made up of the bimetallic strip, onto the strip. The purpose of this screw is to increase the distance that the strip has to be bent in order to open the circuit and turn off the heater. The result is that the screw can be used to adjust the temperature that the heater will turn off at.

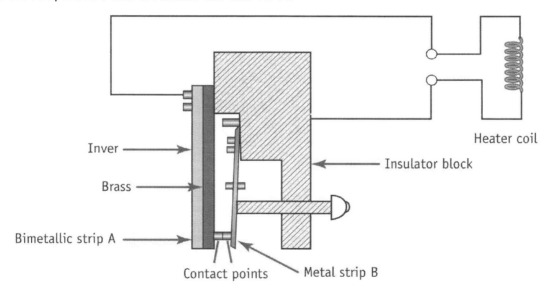

Figure 104 - Thermostat using a bent and flexing bimetallic strip to control the temperature

Problems and solutions to problems with expansion

Buckling in railway lines

Railway lines are long pieces of steel that are laid one to the other. When the temperature rises too far, the railway line can buckle (bend), causing a train to derail. To ensure that this is avoided, the rails are laid with a gap in between so that the rails can expand lengthwise without buckling.

Figure 105 - gaps in rails to allow expansion

Bridges that expand

Most bridges are long enough to expand considerably lengthwise. If this was allowed to happen, the bridge could collapse. As a result, bridges are built that lie on have rollers under one end. As the bridge expands (lengthens) the rollers allow the bridge to move lengthways at one end, while still being stable enough to drive over.

Figure 106 - roller at end of bridge to allow expansion, slots between the joins allow the pieces to slide into and out of each other

Steam and water piping

As hot water goes through a pipe it will expand, and it will expand most lengthways. If this was not controlled, the pipes could buckle, or, if they are cooled, the welding could break. To counter this hoop is often inserted in the pipe that will allow for this movement. When the pipe then expands, or contracts, the hoop enlarges or contracts as a response to the change in the length in the pipe.

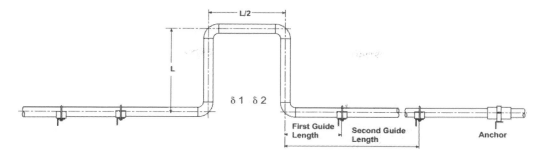

Figure 107 - Hoop in piping to allow expansion

Overhead cables
High voltage cables expand partly as a result of the ambient (surrounding) temperature and also as heat is generated through the flow of current through the cable. If a cable was installed too tightly, when the temperature drops and the cable contracts, it could break. As a result, cables are installed loosely, with the cable drooping between each pylon.

Figure 108 - loose cables on pylons allow for expansion and contraction due to heat

17.8 Expansion of liquids

As with solids, most liquids expand as the get hotter. There is one exception to this; water. We need to recall that density is the amount of matter in a given volume. If the volume increases for a piece of material, its density decreases. The chart below shows this in action for water. Note how the density increases from 0C to 4C, which means that the volume has to be decreasing.

The highest density of water is thus 4°C. Above 4°C, water acts as any other liquid. The reason that this occurs is that as the water changes from a liquid into a solid (ice), it starts to form a lattice, where each bond is longer, thus causing it to expand slightly.

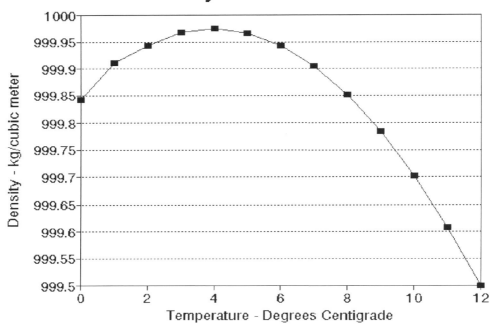

Figure 109 - Graph showing how the density of water is highest at 4C

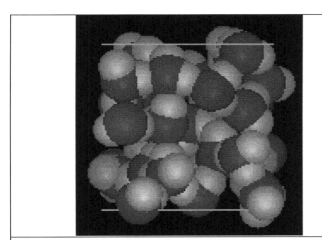	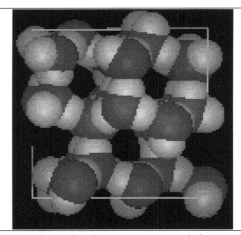
The water molecules in the diagram above as in a liquid state. They nearly fit into the white box.	These water mloecules have rearranged themselves into a lattice structure, which, for water, occupies more space than liquid water, to the left.

17.9 Questions

The diagram below shows a warning system containing a bimetallic strip

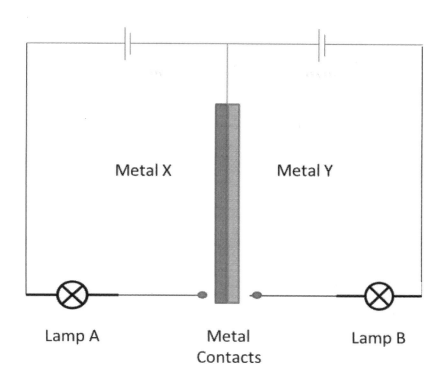

Explain how and why:

1. Lamp B lights up when the temperature increases by 20°C?
2. Lamp A light up when the temperature drops by 20°C?
3. State what the effect would be of moving the metal contacts further away from the strip

Explain the following:

4. A metal bar expands when heated
5. Overhead cables are hung with plenty of slack in them
6. It would *not* be a good idea to reinforce concrete with aluminium rods
7. A bimetal strip bends when heated

Chapter 18. Measurement of Temperature

Although there are many scales which are used to measure temperature with, the two most important ones are C and K (Kelvin). At 0°C, there is still a considerable amount of kinetic energy in the molecules, whereas at 0K, there is no kinetic energy left – it has all been transferred away from the substance. A change of 1°C is the same as 1K, so they are co-linear. However, 0K is -273°C, and 0°C is 273K.

| To convert to kelvin from degrees celsius | Add 273 to the celsius value |
| To convert to degrees celsius from kelvin | Subtract 273 from the kelvin value |

18.1 Temperature

- A measure of the kinetic energy of the molecules.
- Measured in °C or K (kelvin) – not Fahrenheit!!
- Measured using a thermometer

There are several types of thermometers, each using a property of a material or a combination of materials, to show what the temperature is.

- **Liquid in Glass Thermometer**
    - As a liquid heats up, it expands. Two particular liquids might be used; alcohol and mercury. Both expand linearly (see below) as a result of a change in temperature and so the expansion can be measured to see what the temperature is

- **Constant Volume Gas Thermometer**
    - This uses a type of barometer (see Pressure). As the temperature changes, the pressure is shown to change also. Reading off the pressure therefore tells us what the temperature is also

- **Resistance Thermometer**
    - Later on in this book you will be introduced to the concept of resistance changing in a metal when the temperature changes. Measuring and showing the resistance of a material can therefore tell you what its temperature is, if you first calibrate the thermometer properly.

18.2 Liquid in Glass Thermometers

Comparing the two:

Type of liquid	Advantages	Disadvantages
Mercury	<ul><li>Expands evenly when being heated</li><li>Good conductor of heat</li></ul>	<ul><li>Freezes at -39°C</li><li>It is poisonous</li></ul>
Alcohol	<ul><li>Can be used at very low temperatures (freezes at -114°C)</li><li>Expands more than mercury and so can be more sensitive</li></ul>	<ul><li>Cannot be used in hot places</li><li>Has a very low boiling point</li></ul>

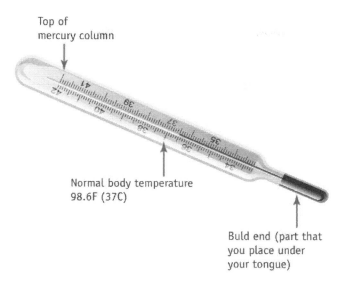

Figure 110 - In glass mercury (or alcohol) thermometer

Key Terms used in thermometers
1) **Linearity** – Even expansion between the two extreme points, the coldest and the hottest.
2) **Sensitivity** – A measure of the expansion of the fluid as a result of a temperature change. Sensitive devices contain a liquid that expands a lot, e.g. alcohol
3) **Range** – the difference between the highest temperature and the lowest temperature that is possible to accurately measure.
4) **Calibration** – the range between the upper fixed and the lower fixed point must be tested. In the case of a thermometer many will insert it in melting water and then in steam, and then put lines between those, representing the 1°C-99°C.

18.3 The Thermocouple

The full inner working of the thermocouple is beyond the scope of iGCSE, but you do need to understand what it does. Two wires of different materials are twisted together and used to probe an area where you are looking for the temperature. One of the wires, additionally, is inserted in iced water, called the "cold junction" and the wires then are connected to a Voltmeter. As the temperature changes, the voltmeter also changes, showing a different potential difference (P.D.). As such you can measure temperature with these by reading the P.D. Most of these need computers in addition to the volt meter to be able to determine what the actual temperature is based on the P.D.

Thermocouples are used to measure:

1) High and low temperatures (beyond the ranges of the liquid in glass thermometers)
2) Rapidly changing temperatures
3) Objects that are distant from the reader. E.g. in a blast furnace or in a fridge.

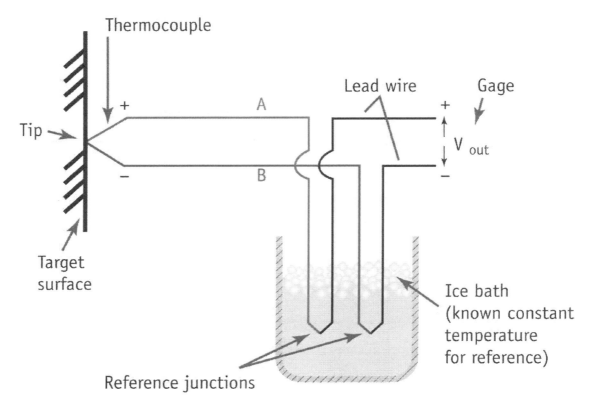

Figure 111 - Typical thermocouple

18.4 Questions

1) State the property used to measure temperature in:
    a. A mercury in-glass thermometer
    b. Thermocouple
2) Give the temperature of the following in Kelvin:
    a. Absolute Zero
    b. Boiling water
    c. Melting ice
    d. 127°C
3) The scale on a mercury thermometer is linear, extending from -10°C to 110°C in 240mm in equal steps:
    a. What is meant by "the scale is linear"?
    b. Calculate the distance move by the liquid if the temperature increases from:
        i. 0.0°C to 1.0°C
        ii. 1.0°C to 100°C
4) Describe how to calibrate this thermometer
5) Describe and explain the features of a thermometer that will make it:
    a. Quick acting
    b. Sensitive
6) Name 3 practical situations where a thermocouple may be used
7) Describe an experiment to accurately calibrate a thermocouple

Chapter 19. The Gas Laws

Gases have more complex behaviours than solids and liquids since they are compressible. As a result there are three variables that must be considered:

1. Volume
2. Pressure
3. Temperature

Before considering how these three things affect each other we need to consider, again, the kinetic molecule model. In gases, where the particles are free to move around, the hotter a gas is, the faster the molecules move. They have a higher kinetic energy. Assuming a volume of gas is enclosed in a container; each molecule will collide with the side and cause it to be pushed. This force is the pressure that the container experiences.

19.1 Boyle's Law (Constant Temperature)

Figure 112 - Robert Boyle 27 January 1627 - 31 December 1691

"For a fixed amount of an ideal gas kept at a fixed temperature, P [pressure] and V [volume] are inversely proportional (while one doubles, the other halves)."

In the mid 1600's, Robert Boyle studied the relationship between the pressure p and the volume V of a confined gas held at a constant temperature. Boyle observed that the product of the pressure and volume are observed to be nearly constant. The product of pressure and volume is exactly a constant for an ideal gas.

$$pV = Constant$$

In analysing this equation, it is found that the units for p x V is Nm (or Joule), and so it is fair to think of this as an extension of the conservation of energy rule, i.e. that the amount of energy at the beginning is the same as the end.

The equation is then:

$$p_1V_1 = p_2V_2$$

Quantity	Name	Units
p_1	Initial pressure	Pa
V_1	Initial Volume	m³
p_2	Final pressure	Pa
V_2	Final Volume	m³

There are two other laws that apply when either the temperature or the volume is constant. They all combine into the ideal gas equation. This, however, is not in any GCSE or iGCSE syllabus. AQAs A2 syllabus covers this in detail though.

19.2 Worked example

A fixed mass of gas at 500Pa (N/m²) with a volume of 30m³ is squashed to a pressure of 1000Pa. Calculate the final volume:

P_1=500Pa
V_1=30m³
P_2=1000Pa
V_2=?

Use:
$$p_1V_1 = p_2V_2$$

Rearrange to make V_2 the focus and we get:

$$\frac{p_1V_1}{p_2} = V_2$$

Then, substituting in the values you get:

$$\frac{500 \times 30}{1000} = 15m^3$$

19.3 Worked Example

A fixed mass of gas at an unknown pressure (Pa) and a volume of 450m³ is compressed to 220Pa and a volume of 150m³. Calculate the initial pressure

P_1=?
V_1=450m³
P_2=220Pa
V_2=150 m³

Use:
$$p_1V_1 = p_2V_2$$

Rearrange to make P_1 the focus and we get:

$$\frac{p_2V_2}{V_1} = p_1$$

Then, substituting in the values you get:

$$\frac{220 \times 150}{450} = 73Pa$$

19.4 Questions

1) A fixed mass if gas at 5000Pa is compressed down to 60m³ and its pressure rises to 260,000Pa, calculate the original volume

2) Calculate the initial pressure of a gas of 393m³ having expanded to 4785m³ at a pressure of 10,000Pa.

3) The pressure inside the gas pipelines is 800kPa. Ten litres of gas escapes into the air where the pressure is 100kPa. What is the volume of the gas now?

4) The final pressure of an escaped gas is 700kPa, occupying a volume of 100m³. If the initial pressure was 2000kPa, calculate the initial volume.

5) Using the molecular kinetic theory, explain why:

   a. Air in a tyre exerts pressure on the walls of the tyre
   b. The pressure inside the tyre increases when the tyre is pumped up
   c. The pressure of the tyre is greater at the end of the journey than at the start

19.5 Gay-Lussacs' Law (aka the Pressure-Temperature Law)

Figure 113 - Joseph Louis Gay-Lussac, 6 December 1778 – 9 May 1850

"The pressure exerted on a container's sides by an ideal gas is proportional to its temperature".

Using mathematical notation this is written as:

$$p \propto T$$

Or, this can be expressed as:

$$\frac{p}{T} = k$$

Where *k* is constant. T must be measured in Kelvin, since this relationship speaks of the kinetic energy of the particles in the container, and the pressure these exert on the wall of the container as a result of colliding with it.

Since, for a given amount of matter, as a gas, in a constant volume, the ratios of the pressure over the Temperature (in Kelvin), is constant, if the pressure, or Temperature, change, the ratio stays the same.

This can be expressed as follows:

$$\frac{p_1}{T_1} = \frac{p_2}{T_2}$$

Quantity	Name	Units
p_1	Initial pressure	Pa
T_1	Initial Temperature	K
p_2	Final pressure	Pa
T_2	Final Temperature	K

Chapter 20. Specific Heat Capacity

When an object is heated, its particles, at a microscopic level, gain more internal energy – a combination of kinetic and potential energies. This requires the transfer of energy into the material. Each material responds differently to this transfer of energy and rises in temperature different amounts as a response to the new energy. The material constant that describes how much energy is required to heat up 1 kg of the material by 1 degree Celsius (or Kelvin) is called the specific heat capacity, and it is measured in J/Kg°C.

The relationship between heat and temperature change is usually expressed in the form shown below where c is the **specific heat capacity**.

$$E = cm\Delta T$$

Quantity	Name	Units
E	Energy required (or released)	J
c	Specific heat capacity of the material	J/Kg°C or J/KgK
m	Mass of material being heated, or cooling	Kg
ΔT	Change in temperature	°C or K

20.1 Worked example

If 2Kg of water cools from 70°C to 20°C, how much energy is released into the surroundings?

ΔT= 70-20=50°C
m=2Kg
c=4200J/Kg°C

Therefore E=4200x2x50=420000J

Compound	Specific Heat Capacity (J/Kg°C)
Water	4200
Alcohol	2500
Ice	2100
Glass	700
Copper	400

20.2 Thermal Capacity

When we look at the specific heat in more detail we can see that we can also define **thermal capacity** as a useful way to think of how much energy is required to heat an object 1 degree (°C or K) :

	Quantity	Name	Units
$$thermal\ capacity = cm$$	Thermal capacity	Thermal capacity	J/°C or J/K
	c	Specific heat capacity of the material	J/Kg°C or J/KgK
	m	Mass of material being heated, or cooling	Kg

Therefore we see that thermal capacity is measured in J/°C – or the amount of energy per degree raised (or dropped).

The quantity, thermal capacity * ΔT, then becomes the amount of energy required (or released) when change the temperature by ΔT degrees (either Kelvin or Celsius).

20.3 Power and Energy

Many exam questions will ask how much energy is transferred to an object through a heater – here it useful to remember:

$$Power\ x\ time = Energy$$

Or

$$E = Pt$$

For example, if a 2kW heater is used for 1 hour, how much energy has it transferred?

A: 2000W x 3600s *(1 hour)* = 7.2MJ

Experiment to determine the specific heat capacity of water

In the following practical, a mass of water is heated (1kg), using a 100W heater, for 180s. The start and end temperatures are recorded. This experiment is run "as is". In analysing it, it is found that the heat from the the thermometer takes time to reach the thermometer, so the change in temperature change is smaller than it should be. The immersion heater does not either transfer all its energy to the water, and some of the heat is lost to the surroundings, even with the insulation. So the value of the specific heat capacity can only be indicative, not accurate.

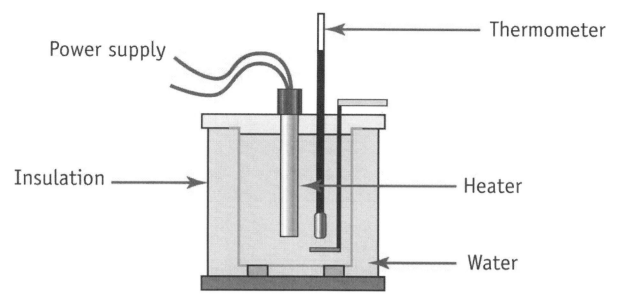

Figure 114 - Measuring specific heat capacity of water

1) use the setup above
2) add 1Kg of water (approximately 1l)
3) have a 100W heater
4) measure start temperature
5) set clock going for 3 minutes (180s)
6) measure end temperature

Using the specific heat capacity equation:

$$E = cm\Delta T$$

And reorganizing to make c the subject, we have:

$$\frac{E}{m\Delta T} = c$$

Substituting in the values:

E=100W x 180s = 18000J
m=1Kg
ΔT=the measured change (in our experiment with water, this change was 4.5°C)

Therefore we would find c as:

$$\frac{18000}{1 x 4.5} = c = 4000 J/Kg°C$$

In iGCSE questions you are sometimes asked what assumptions you have made, and in this case we have assumed that all the energy supplied by the heater was heating the water. In reality some of it was heating the surroundings also.

Finding other unknown entities in the specific heat capacity equation
The equation consists of 4 different entities, and finding any one of them involves rearranging the equation:

$$E = cm\Delta T$$

20.4 Worked example

Assume that 4Kg of water is being heated, and its temperature is increased from 30C to 85C, and the specific heat capacity of water is 4200J/Kg°C, how much energy was transformed to heat?

c=4200J/Kg°C
m=4Kg
ΔT=85-30=55°C

Therefore the energy transformed was = 4200x4x55=924kJ

20.5 Questions

1) Calculate the energy needed to heat 20kg of water from 15°C to 90°C
2) Calculate the energy lost by a 5kg block of aluminium when it cools from 100°C to 40°C
3) A 220W heater is placed in 4kg of liquid for 2 minutes:
    a. Calculate the energy transferred by the heater
    b. Calculate the specific heat capacity of the liquid if it rises from 20°C to 45°C
4) A 2kg block of iron is given 10kJ of energy and its temperature rises by 10°C. Calculate the specific heat capacity of the iron.
5) Calculate the energy needed to change the temperature of:
    a. 2kg of water by 5K
    b. 500g of water by 4K
    c. 100g of gold from 20°C to 30°C
    d. 200g of copper from 60°C to 10°C
6) How much energy is needed to heat 100,000 kg of water in a swimming pool from 15°C to 20°C
7) Calculate the energy transferred into 500g of water when heated from 20°C to 100°C
8) The specific heat capacity of concrete is 1050J/kg°C. It is part of a heater. If the heater cools from 80°C to 40°C, how much heat energy is given out from a 25kg block of concrete?
9) A 60W heater is switched on for 90 seconds. It is used to heat 2kg of a liquid. The temperature goes from 23°C to 25°C. Calculate the specific heat capacity of the liquid.
10) Calculate the energy lost from a metal X (specific heat capacity=800J/kg°C) if its mass in 800g and it cools from 70°C to 30°C.

Chapter 21. Latent Heat

There are three **phases** or **states** of matter, such as water:

1) Solid (Ice)
2) Liquid (water)
3) Gas (steam)

As the material changes from one phase to another it either absorbs, or releases, heat.

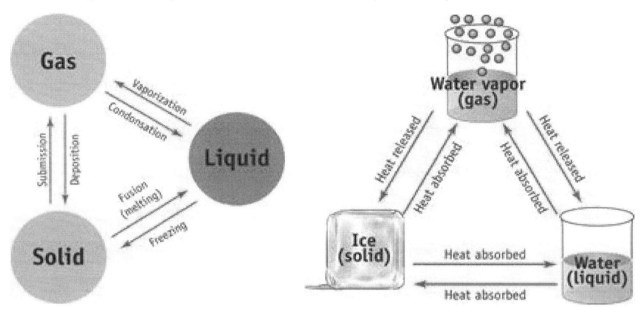

Figure 115 - How heat is transferred between the states

Whilst in a particular phase, this heat transfer increases the temperature, but between changes in phases, e.g. from a solid to a liquid (fusing), this absorbed heat is used to separate the particles to form the liquid. The energy used to separate the particles from a solid phase to a liquid phase is called the **specific latent heat of fusion**.

21.1 Specific latent heat of fusion (or freezing)

The relationship between the specific latent heat of fusion and the amount of energy required to melt (or freeze) is:

	Quantity	Name	Units
$$E = mL$$	E	Energy required or released	J
	L	Specific latent heat	J/Kg
	m	Mass of material being heated, or cooling	Kg

where L is the **specific latent heat**. For water, this specific latent heat is 330000J/Kg. This means that it would require that 330000J of energy is absorbed per Kg of water to change its phase to (or from) a liquid.

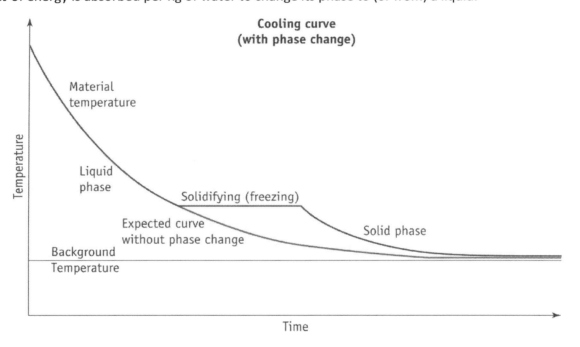

Figure 116 - A cooling curve - the temperature drops until it changes phases and then levels off until it has completely solidified. The temperature then continues to drop.

The diagram above shows how the temperature of a melting substance changes over time. As it fuses (becomes a solid), the temperature stays constant for a period of time. When the substance has become a solid, its temperature continues to drop. The diagram also shows what would have happened if there was no phase change, and thus no specific heat of fusion.

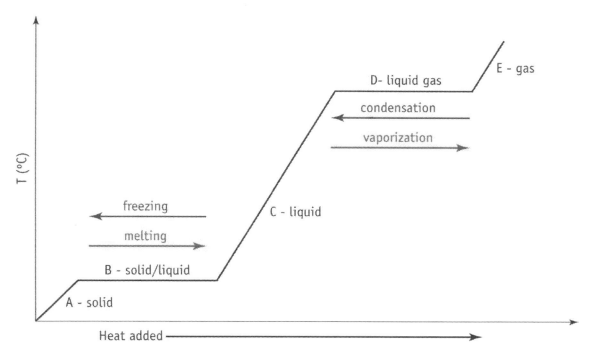

Figure 117 - Phase change diagram, showing the stages where latent heat is used to change the phases

Figure 117 - Phase change diagram, showing the stages where latent heat is used to change the phases shows a theoretical phase change diagram for an idealized substance (in reality, these graphs are never as neat as this).

21.2 Specific Latent Heat of Vaporization

When the phase change is from a liquid to a gas, the latent heat is called the **latent heat of vaporization**. When boiling water, the water will not rise beyond 100°C, but will continue to boil and change phase to a gas (steam). It does so through the absorption of energy from a heater. The energy required to change it to a gas is used to further separate the particles so they can form a gas. For water this is 2300000J/Kg. The same equation as above is used to calculate that amount of energy absorbed (or released, if condensing).

Performing experiments to determine specific heats

Measuring specific heat of fusion
At iGCSE this experiment is done without the control, but it would end up giving a specific heat of fusion that is too low. In the experiment outlined below, you will have one experiment running with an immersion heater that is switched on, and one with an immersion heater switched off. Apart from that difference, the experiments should be set up exactly the same. This ensures that the control variables are considered.

Place the two immersion heaters in the funnels and pack the crushed ice around them. The element of the heaters should be positioned towards the end of the metal barrel; check that this is in good contact with the ice. Connect one heater to the power supply but do not turn it on.

1) Record the weights of the two beakers and place them under the funnels.
2) Wait until both funnels are dripping water into their beakers then turn on the power supply and at the same time start the stop clock.
3) Record the current and potential difference of the supply circuit and do not alter it. (remember that Pt=E and P=VI, so E=IVt)
4) Ensure that the ice is kept topped up and in good contact with the two heaters.

5) Run the experiment for 5 minutes
6) Turn off the power and stop the clock but do not move the beakers.
7) After another minute or so (when the flow of melt water is back to the rate of the control setup) remove the beakers and weigh them again

This will give you the amount of water melted due to the heater (if you use the control, you should subtract the mass of the control beakers' contents, which has been melted due to the room temperature).

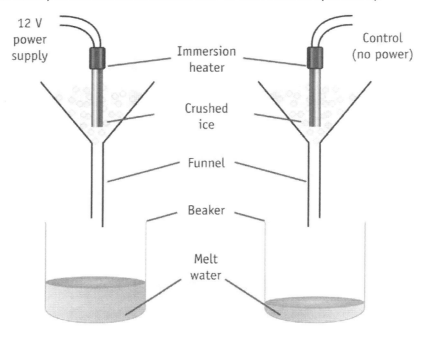

Figure 118 - Experiment to find the latent heat of melting of ice to water

The equation to use is:

$$E = mL$$

Which we will rearrange to change the focus to L:

$$\frac{E}{m} = L$$

If we substitute in the required relationship for the energy supplied, E, we get:

So if we supplied 12V at 5A for 5 minutes (300s) for a mass of 70g of water we get L to be:

$$\frac{12 \times 5 \times 300}{0.07} = L = 257000 J/Kg$$

Measuring specific latent heat of boiling
This process involves boiling a beaker of water using an immersion heater and then measuring how much mass the beaker has lost as the water has boiled off.

Figure 119 - measuring specific latent heat of boiling

1) Put 100ml of boiling water in a beaker and weigh the combination of beaker and water – record this as m_1
2) Insert the immersion heater in the water and turn on the power
3) Let the water boil for 5 minutes
4) Remove the immersion heater and weigh the beaker and water again, and record as m_2

Use the following information in the

$$\frac{E}{m} = L$$

Here E will be VIt (voltage x current x time) and the m will be the difference between m_1 and m_2. Using this information it is possible to calculate the **specific latent heat of boiling**.

21.3 Questions

1) If the specific latent heat of ice is 340,000J/kg, calculate:
    a. The energy needed to melt 2kg of Ice at 0°C
    b. The energy needed to melt 500g of ice at 0°C
2) Calculate the latent heat of fusion if 260,000J of energy is used to melt 90°C

3) The graph shows a cooling curve of substance as its temperature falls from 300°C and 20°C

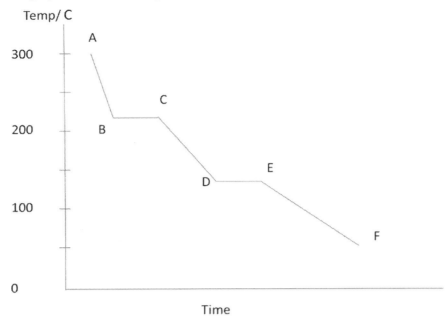

   a. At 250°C is the substance a solid, liquid or gas?
   b. What is the boiling point of the substance?
   c. What is the melting point of the substance?
   d. Where is there a transfer of energy without a decrease in temperature?
4) A heater of 60W is switched on for 5 minutes and melts crushed ice. Calculate the latent heat of fusion of ice if 40cm³ of water is collected
5) Calculate the energy needed to melt 4g of a substance with a specific latent heat of fusion of 540,000J/kg
6) Calculate the specific latent heat of vaporisation is a 2000W kettle boils 1200g of water to 1100g in 8 minutes
7) Calculate the energy transferred if 3kg of water is heated (the specific latent heat of vaporisation of water is 2.3MJ/kg)
8) An electric kettle rated at 3kW heats 600g of water for 12 minutes. The mass changes to 450g. Calculate the specific latent heat of vaporisation of the water.
9) Calculate the energy needed to boil 100g of water at 100°C (the specific latent heat of vaporisation of water is 2.3MJ/kg)

Chapter 22. Heat Transfer

When a material is being heated, such as a rod of metal, the heat will gradually move along it from the heat source towards the colder end.

The speed at which this heat is transferred depends on:

1) The material used – metals are the best conductors of heat
2) The difference in temperature between the hot end and the cold end

The method of heat transfer varies between the different phases, solid, liquid and gas. Liquids and gases have a common method of heat transfer and so we will use the generic name for the two, a **fluid.** Liquids and gases are both fluids.

There are three main methods of heat transfer

1) Conduction – in **solids**
2) Convection – in **fluids**
3) Radiation – from hot bodies, using infrared radiation

22.1 Conduction

As a solid is being heated, there are two ways that the heat is transferred from the hot end to the cold end.

1) Through particles bumping in to each other. When a particle in a solid heats up with start to vibrate more, taking up more space and it will start to bump against its neighbours. As it does so, it transfers some of its energy (heat) to them. As a result the also become hotter.

2) Specifically in **electrically conducting metals**, such as copper or brass, there is a "sea" of free, delocalized, electrons that are able to move around (these are called valence electrons). When energy is transferred to these they start to wander away from the heat, taking the extra energy with them, and, as they bump into other particles, they transfer this to them, heating them up. This process is much quicker than the first method, and so good electrical conductors are also good heat conductors.

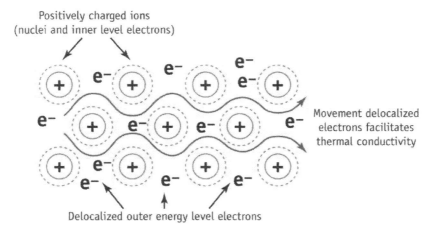

Figure 120 - Heat transfer in electrically conducting metals via electron movement

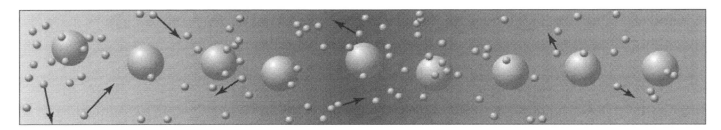

Figure 121 - How a "sea" of electrons moves heat away from the hot area

Good Conductors/Poor Insulators	Poor conductors/Good insulators
Metals such as: Aluminium, Copper and Brass	Plastics
Silicon	Glass
Graphite	Wood
	Wool
	Fibreglass
	EPS

Insulators, such a fiberglass and special types of EPS (expanded polystyrene) are used to insulate houses, for instance, to ensure that heat is kept controlled – inside in winter, and outside in summer. Carpets also help in insulating buildings from conduction of heat to or from the ground.

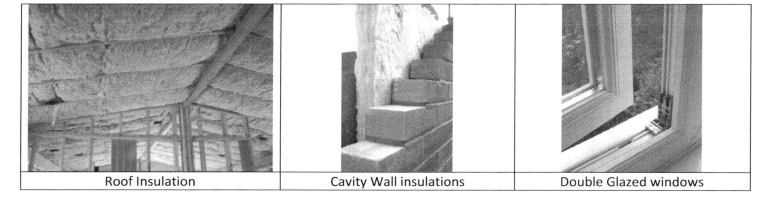

| Roof Insulation | Cavity Wall insulations | Double Glazed windows |

Demonstrations of conduction

In the picture beside, you can see a star formed out of different types of metal rods. At the end of each rod there is a small square of paraffin. A flame from a Bunsen burner is placed under the middle of the star. This will heat the rods and the heat will transfer out, away from the centre toward the paraffin. When the paraffin is hot, it will melt and fall into the tray. This demonstrates how heat is conducted differently in different metals.

Figure 122 - Practical to illustrate different rates of conduction

22.2 Convection

Since the particles in fluids (liquids and gases) are are not fixed to each other, the method of transferring heat around is different than in solids.

1) Gases are BAD conductors since the particles are too far apart to bump into each other regularly, or for electrons to transfer the energy
2) The particles in fluids are free to circulate
3) When free particles carry heat energy it is called **convection**

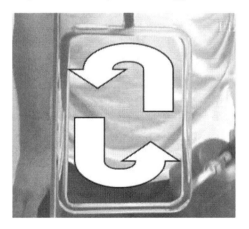

In the two demonstrations above you can see how liquids and gases behave the same:

1) Particles that are heated rise
   a. In the demonstration on the right, the air is heated by the candle at the bottom.
   b. In the demonstration on the left, the water is caused to rise by the gas burner at the bottom right
2) As the particles rise, they are replaced by cooler ones
   a. On the right, the air is drawn in from the left into the cabinet and then out again through the right tube with the rising air
   b. On the left, the water in the pipe is replaced by water from the rest of the pipe.

Both demonstrations show how **convection current** is set up by the movement of hotter fluids upwards and then being replaced by the dropping, cooler fluids.

Examples of convection in action

Day and night-time winds
Convections occur in larger world; for instance in the movement of on-shore and off-shore winds by the sea-side. During the day the sun heats up the land more rapidly than the sea and so the warm air rises as it is displaced by the cooler air moving in from the sea.

The lower part of the diagram shows how, at night, the sea, which cools less than the land, becomes the warmer region and so the warmer air there rises as it is displaced by the cooler air moving out from land.

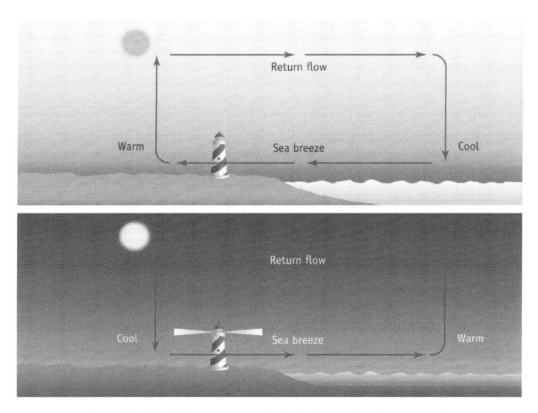

Figure 123 - Winds blow onto land during the day, and out to sea at night

Room heating
As a heating element warms the air, it rises toward the ceiling, and is displaced by the cooler, denser, air that is cooled by the window. This causes a convection current around the room.

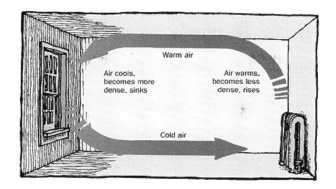

Figure 124 - convection currents carry the heat up, across and down a room from a heating element

Freezer compartments

These compartments are placed highest up in the fridges and cools the air down, causing it to become denser as the particles lose energy. This displaces the warmer air, which rises up to the freezer, thus becoming cooler. This becomes a convection current around the inside of the fridge.

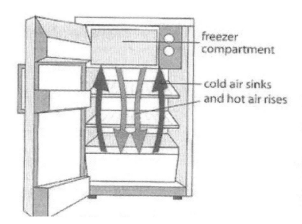

Figure 125 - Cold, more dense, air from the freezer falls, displacing the warmer air up to the freezer compartment

Hot Water System

In a home, the water is heated in a boiler and is therefore caused to rise. This water rises to the upstairs, where it is used in the radiators, cooling as it passes through the heaters. The water then becomes denser and drops down to the lower floor and continues to be used in the radiators, further cooling. A convection current has been created by the boiler that will continue to circulate the water around the building.

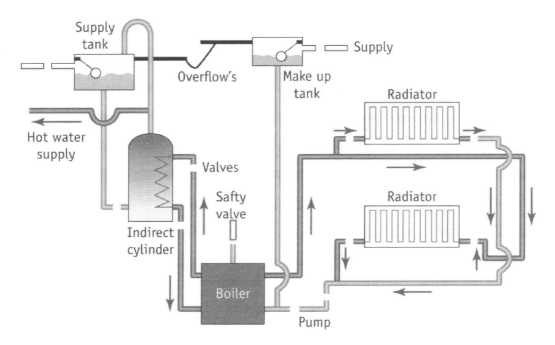

Figure 126 - Hot water heating using convection

22.3 Radiation

The last of the three methods of heat transfer uses information that we will learn more about, namely a form of light that we cannot normally see: **Infrared Radiation**

Figure 127 - Heat radiating from a cup via infrared radiation

1) Infrared Radiation is similar to light
    a. Part of the Electromagnetic Spectrum (see later)
    b. All hot items emit **thermal radiation**
2) Very hot items also emit visible light – e.g. a red-hot poker
3) The intensity of the radiation depends on:
    a. The type of surface of the emitter
    b. The temperature of the emitter

Emitters and absorbers

Good emitters of thermal radiation are also the best absorbers, as the diagram shows below.

- Black colour is the best emitter, silver is the worst.
- Silver is the worst absorber

The finish of the material is also important:

- Shiny silver is the worst emitter and the worst absorber
- Matt is the best emitter and the best absorber

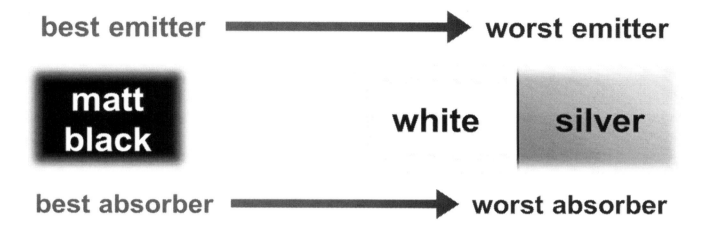

Figure 128 - Matt black is both the best radiator and absorber of radiated heat

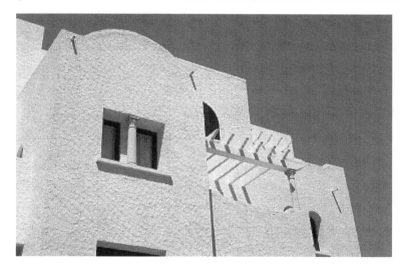

Figure 129 - Houses in hot countries are painted white to reflect the heat

So, the best absorber and emitter combination will be a matt, black surface. In most hot countries, the houses are built and painted white to stop the heat from being absorbed and keeping the houses cooler.

The Vacuum flask

A vacuum flask uses all the available methods of keeping heat from escaping.

- Usually the outside of the flask is a shiny silver finish – this reduces the thermal radiation from the flask
- The liquid inside is separated from the outside of the flask by a gap, thus stopping conduction of heat through the material.
- The inner bottle and the outer bottle are separated by a vacuum, thus stopping the flow of het through convection.
- The flask has a screw top to stop convection from occurring at an opening
- The support at the bottom is insulated

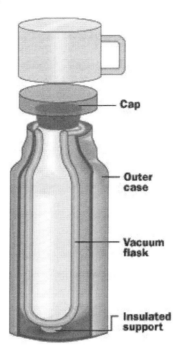

Figure 130 - Vacuum flasks use many techniques to keep the heat in

The solar panel

Solar panels are often added to modern houses to heat water for "free" from the radiation that the Sun emits. These are best when they have large surface areas and are made black, so that they absorb as much of the radiation as possible.

Thermography

Since warm things give off heat, being able to detect heat with a camera and varying degrees of heat can be useful, especially in medicine, where it can detect where blood flow is poor, or in joints that are damaged and hotter than usual.

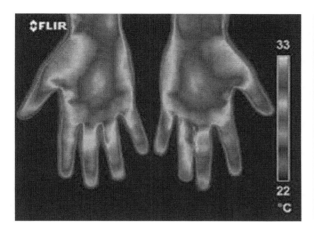

Figure 131 - Thermography used in medicine and in checking house insulation

The Green house

A greenhouse is designed to let in the radiation from the sun whilst not letting it out. The plants absorb the heat, but they emit their radiation at a longer frequency, which cannot pass through the glass.

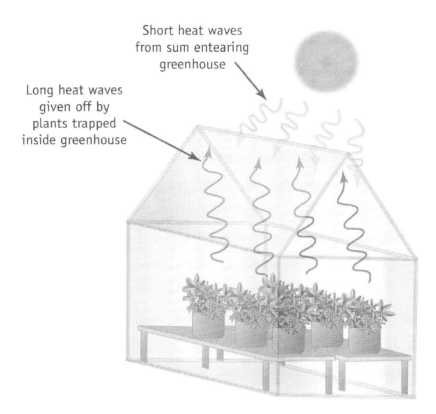

Figure 132 - Greenhouses allow infrared light to enter, but not to escape

Volcanologists and Foundry workers
People who work near very hot heat sources need to ensure that they stay safe. They therefore wear protective clothing with a silvery surface that reflects the heat.

22.4 Questions

1) Describe three ways that heat loss in a house maybe reduced. Give:
    a. Two features of the vacuum flask that prevents heat loss by conduction
    b. One feature of the vacuum flask that prevents heat loss by convection
    c. One feature of the vacuum flask that prevents heat loss by radiation

2) Explain:
   a. Why saucepans have copper bottoms and plastic handles
   b. Wool and feathers are good insulators
   c. It is safer to pick up a hot dish with a dry cloth than a wet one
   d. Why are metals much better thermal conductors than most other materials
   e. A hot water tank loses thermal energy even when lagged. How could the energy loss be reduced?
3) Explain:
   a. A radiator warms the air in a room even though air is a poor thermal conductor
   b. The freezer compartment in a refrigerator is placed at the top
   c. Smoke from a bonfire rises upwards
4) On a hot summer's day, coastal winds blow in from the sea.
   a. What causes these winds?
   b. Why do they change direction at night?
5) Name the surface that is best at:
   a. Absorbing thermal radiation
   b. Emitting thermal radiation
   c. Reflecting thermal radiation
6) Explain:
   a. Why polystyrene is used for drinking cups
   b. Why a lid with a hole covers a polystyrene cup

Chapter 23. Common Wave Terminology

23.1 Background

- Waves are caused by oscillations, or vibrations.
- Waves transmit energy from one place to another
- Waves repeat very specific patterns

23.2 Wave Terminology

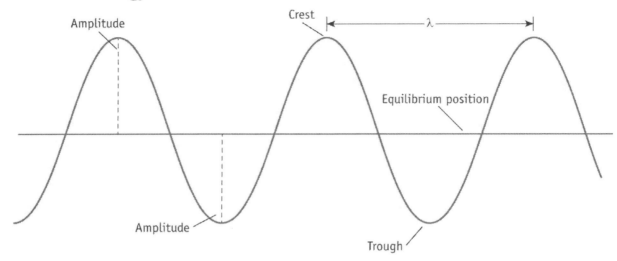

Figure 133 - Wave terminology

- The **equilibrium position (or rest position)** is the position that a wave would end up if ceased oscillating
- The **wavelength** is the distance between two points on a wave that has the same **displacement**. As we will see, this also needs to be when the wave is also moving in the same direction (up or down). As this is a length, this is measured in metres. The symbol used for this is λ, lambda.
- The **displacement** is how far from the **equilibrium position** (or rest position) the wave is. This value can be positive or negative
- **Crests** (or **Peaks**) are the highest point, the one with the largest **displacement**, in the **positive** direction
- **Troughs** (or **valleys**) are the lowest point on a wave with the largest **negative displacement**
- The **amplitude** is the largest displacement in either direction.
- **Frequency** is the number of oscillations, or waves that passes a point every second and is measured in Hz (Hertz), or s^{-1}.
- **Time Period** – The time it takes to for one oscillation to take place

Examples of measuring wavelengths can be seen in figures 2 and 3, below. Note how it is important to locate the same exact spot on each wave, where it is at the same displacement AND where is it going up (or down) in the same direction as the previous wave.

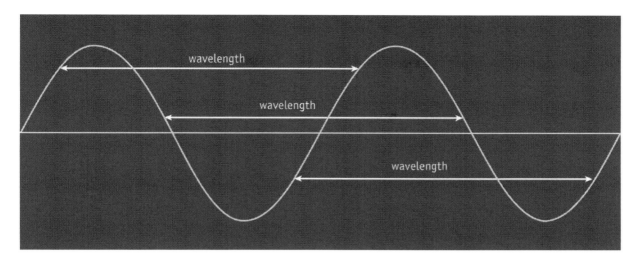

Figure 134 - Wavelengths measured as various parts of the wave

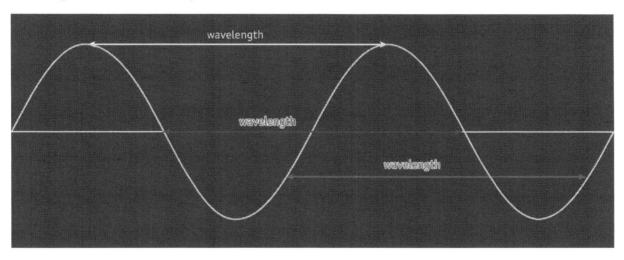

Figure 135 - More typical places to measure wavelengths

23.3 Time Period

When discussing pendulums above we noted that the time period is related to the frequency by the following equation:

$$T = \frac{1}{f}$$	Quantity	Name	Units
	T	Period of one oscillation	s
	f	Frequency	1/s or Hz

23.4 The Wave Equation

It turns out that it is very easy to calculate the velocity of the propagation of the wave (v) based on the wavelength and the frequency. This equation is called the Wave Equation:

Velocity = frequency x wavelength

Or

$$v = f\lambda$$	Quantity	Name	Units
	v	Velocity of the wave	m/s
	f	Frequency	1/s or Hz
	λ	Wavelength	m

In this equation, v is the velocity; f is the frequency and λ is the wavelength. As with all physics, it is important to remember to use the correct units (m/s, s^{-1}, and m respectively).

Example:

A wave has the frequency of 100Hz and the wavelength of 5m, what is its velocity?

$$v = 100 \times 5 = 500 m/s$$

23.5 Types of waves

There are two types of waves that you need to know at iGCSE level:

1) Longitudinal waves (e.g. sounds waves or a slinky being pushed forwards and pulled backwards)

2) Transverse waves (e.g. electromagnetic radiation such as light, water waves and earth quakes)

Longitudinal waves
In longitudinal waves, the oscillations occur forwards and backwards in the direction of the movement. In the case of sound, this oscillation involves individual particles moving to and fro, into and then out of, compressions.

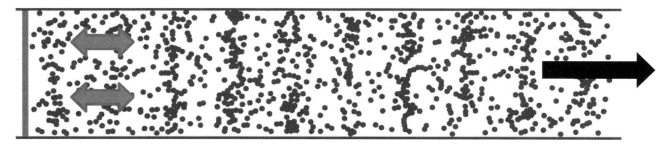

Figure 136 - Compressions in a gas with the individual particles only move from left to right. The wave travels to the right.

The areas where the particles are gathered together are called "compressions" and the areas in between, where the particles are scarcer, are called "rarefactions".

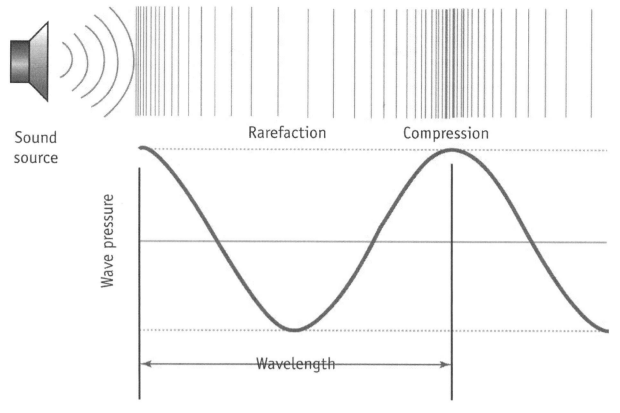

Figure 137 - Rarefactions and compressions in a sound wave

The diagram shows a compression and a rarefaction in a sound wave, but the same is also true for a slinky that is pushed and pulled to form a longitudinal wave:

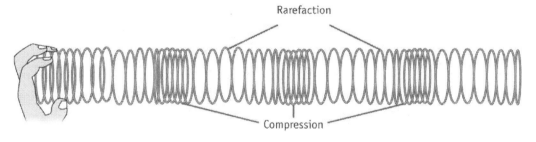

Figure 138 - Rarefaction and compression in a slinky

Longitudinal Waves
These waves oscillate at 90° to the direction of movement, from side to side. These waves only act transversely.

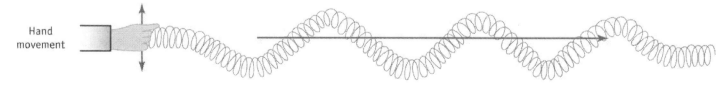

Figure 139 - Slinky being shaken from side to side to generate a transverse wave

Figure 140 - Water waves, where the varying height of the wave shows the up and down motion of the oscillations

23.6 Normals

Waves are best understood using angles and these are ALWAYS used relative to the "normal". A normal is an imaginary line that is perpendicular (at 90° to) the surface that the wave is interacting with. The diagram below show the simplest form of finding the normal:

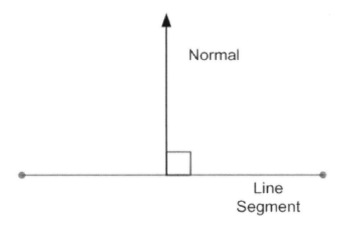

Figure 141 - Normal to a straight line segment

Sometimes, however, it is vital to find the normal to a curve instead of a flat line. In this case it is done in two stages:

1) Draw a tangent to the curve at the point of interception

2) Draw the normal to the tangent at the point of interception

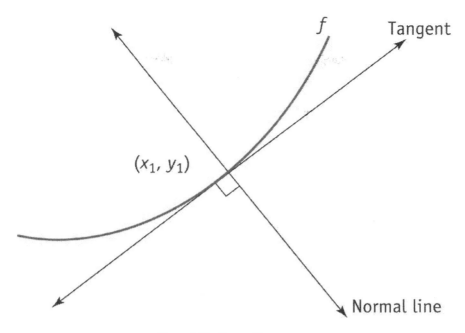
Figure 142 - Normal to a curve.

As we will see, getting this right is essential for all wave interactions.

Chapter 24. Wave Effects

24.1 Ripple Tanks

Ripple tanks can be used to generate waves and show how waves interact with objects and varying depths. A machine is set to oscillate up and down with either a small ball, or a piece of wood slightly submerged in water. The result is a set of waves that will move across the table. Ripple tanks usually have a glass bottom and a light at the top which causes a play of the waves to be displayed under the tank, on the floor.

Figure 143 - Ripple tank

There are certain phenomena or effects that show that waves are present and that determine how waves respond to certain environmental changes, such as objects that block their path.

24.2 Reflection

When waves interact with a plane surface it will reflect back off it, as shown below.

- The wave angle of the wave coming into the object is the same relative to the "normal".
- The speed, frequency and wavelength stay the same

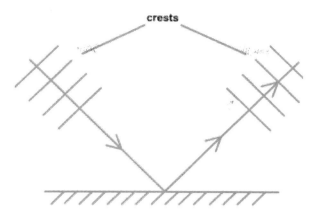

Figure 144 - An incident ray, reflecting off a surface.

If we draw in the normal to this we can see how two very important angles are defined:

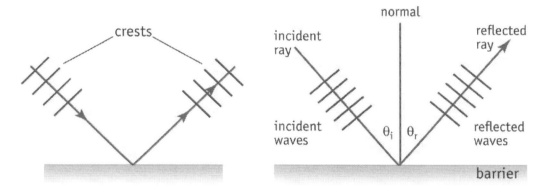

Figure 145 - The two angles are called the "incident angle" and the "reflected angle" they are often represented by the Greek letter, Theta.

When a wave is reflected off a curved surface it is simple to determine how continue their path using the concept of adding a normal to each point where the ray intercepts the barrier. In the case of a concave mirror (where the mirror is curved inwards), the rays converge to one point. In the case of a convex mirror (one that curves outwards) the rays will spread away from the barrier. The convex mirror appears to have an imaginary (or virtual) point behind them where the rays appear to emanate from:

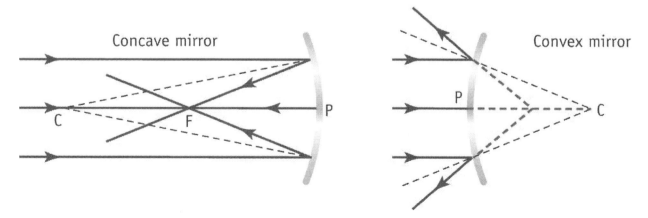

Figure 146 - Concave and convex mirrors reflecting rays

The following diagram shows how wave fronts affect these mirrors differently.

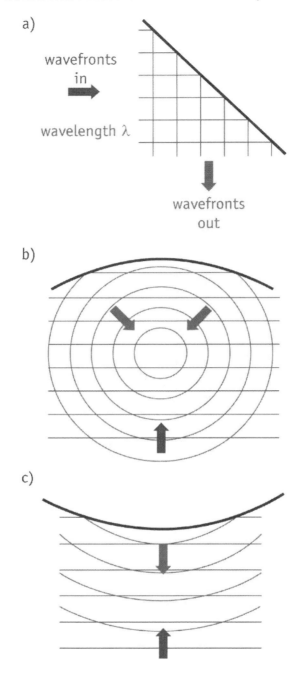

Figure 147 - Wave front interaction with barriers, such as mirrors

24.3 Refraction

We will initially explore the phenomenon of refraction in water waves and then further with light. As a wave moves through water of a particular depth it keeps the same speed, wavelength and frequency. However, when it encounters shallower water, the speed of the wave slows down (the reverse is true also – when it moves from shallow to deep, it speeds up). As a result, the wavelength shortens. The frequency may not change in a wave as a result of any refraction.

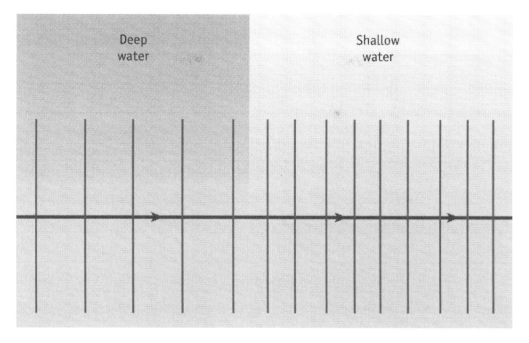

Figure 148 - Speed change as a wave moves through different depths of water

It is possible to see this effect when waves approach land and the waves become closer together.

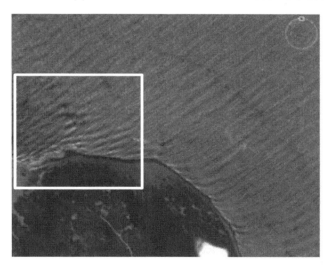

Figure 149 - Note how the waves change both direction and wavelength as the approach land

As the photograph above shows, waves not only change their wavelength, but, as a result of changing their speed and wavelength, they also change their direction. We will show more about this later. The area in the box above is shown, diagrammatically, below.

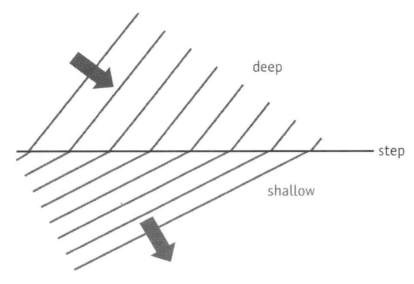

Figure 150 - Refraction in water, showing change of direction and wavelength

24.4 Diffraction

When a wave encounters an obstacle that forces it through an opening, the waves will diffract. The way they do this depends on the width of the opening compared to the wavelength of the wave. If the opening is more than the wavelength, the waves will propagate with a slight bend at each end:

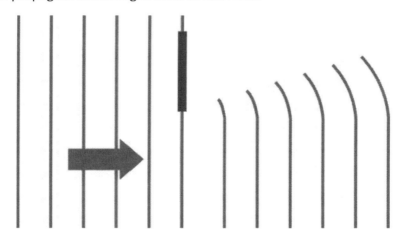

Figure 151 - Diffraction at an edge, or when passing through a wide opening

- Only bend at the edges
- Waves travel straight on
    - Known as rectilinear propagation

If, however, the gap is about the width of a wavelength the wave will diffract as a circular waveform:

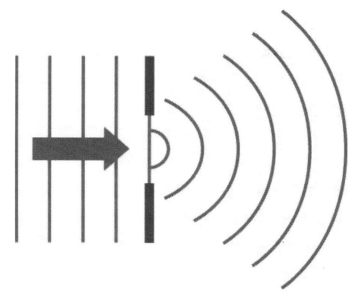

Figure 152 - Diffraction though and opening where the width is about the size of the wavelength

- Curved waves
- Gap about the same as the wavelength

Figure 153 - Diffraction at an inlet

24.5 Questions

1. A wave transfers _____ in the direction that the wave travels. A transverse wave is a wave where the vibrations are _____ and a longitudinal wave is a wave where the vibrations are _____
2. A cork floats on the surface of water on a lake. A series of small waves pass by. Describe the motion of the cork.
3. What is meant by the frequency of a wave?
4. The diagram in Figure 1 shows a wave. Mark on the diagram (a) the amplitude and (b) the wavelength.

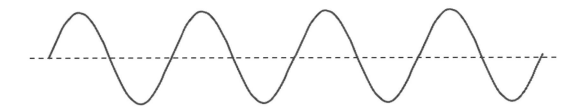

5. If the pitch of a note is increased what happens to:
    a. the frequency of the note
    b. the wavelength of the note
6. a wave has a frequency of 100 Hz and a wavelength of 3 m. Calculate the speed of the wave.
7. a wave moving at 40 m/s has a wavelength of 2 m. What is its frequency?
8. Two ripples move across the surface of a pond. One has a wavelength half that of the other. What can you say about the speeds of the two waves?
9. Complete the wave diagram below, where a flat wave approaches a plane barrier at 45° and reflects.

10. Complete the diagram to show what happens to the waves at positions A, B and C

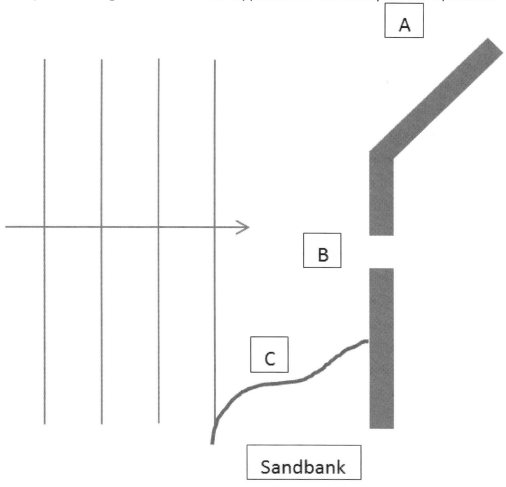

Chapter 25. Light

Light can be formed through two processes:

- Incandescence is the emission of light from "hot" matter (T ≳ 800 K). For instance the filament in a light bulb.
- Luminescence is the emission of light when excited electrons fall to lower energy levels. For instance low energy light bulbs.

All other objects, such as the Moon and the smoke particles in the Brownian motion experiment, reflect the light.

25.1 Properties of light

- Light is a part of the electromagnetic spectrum (see further down)
- Light transfers energy
- It is made up of waves, and is also a particle, called a photon (the photon is covered in AS-Physics)
- It always travels in straight lines
- Able to travel in a vacuum
- Travels at 3×10^8 m/s in a vacuum
- Can produce shadows – formed when opaque objects block the propagation of light
- Undergoes all the wave effects of reflection, refraction, diffraction and, additionally, dispersion

25.2 Reflection

- Occurs when light bounces off a surface (theorists believe that the surface absorbs the light ray and then re-emits it according to the rules of reflection, set out above)
- There are two types of reflection:
1) Regular (this is the main one that you will learn about, but is an idealised version that is seldom possible in reality)
    a) Regular reflection has all incident and reflected rays obeying the reflection rule, set out above. All rays are parallel on the way in and on the way out.
2) Diffuse (this is the one we observe mostly and that can sometimes cause confusion when trying to observe Total Internal Reflection and other effects covered later)
    a) ll incoming rays are parallel, but reflected rays are going in all directions and are typically not parallel. When observing Total Internal Reflection (see later) this can lead to unexpected early reflection.

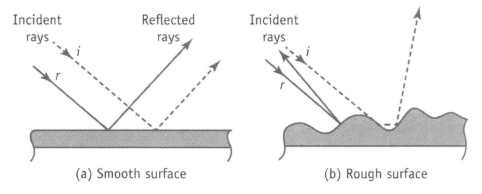

Figure 154 - the two types of reflection of light

25.3 Reflection of light on a plane mirror

- When a regular reflection is achieved, the incoming and reflected angles are the same relative to the Normal.
- The angle between the incident (incoming) ray and the normal is called the incident angle
- The angle between the reflected (outgoing) ray and the normal is called the reflected angle

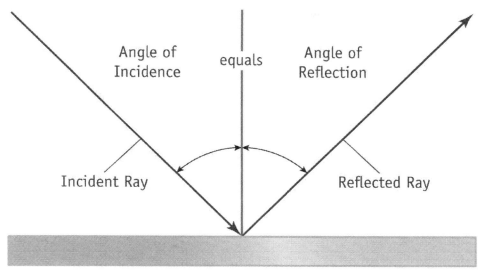

Figure 155 - First law of Reflection - angle of incidence = angle of reflection

	Quantity	Name	Units
$incident\ angle = reflected\ angle$	Reflected angle	r	Degrees
	Incident angle	i	degrees

Laws of Reflection

1) The Angle of Incidence = the Angle of reflection
2) The incident ray and the reflected ray, together with the normal, are in the same plane (they create a flat surface)

Figure 156 - Reflection in a mountain lake

Reflection of an image in a plane mirror

When images are seen in mirrors they have certain properties, assuming the mirror is plane with the object being observed:

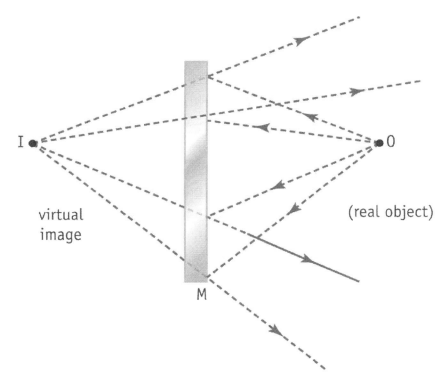

Figure 157 - The location of a virtual image in a plane mirror

1) The image is upright
2) It stays the same size

3) It is the same distance from the mirror, but appears behind it
4) It is laterally inverted, right and left are swapped over
5) It is virtual, i.e. it appears real but is behind the mirror in a mirror world
6) The line joining the object to the mirror is perpendicular (at 90°) to the mirror

Figure 158 - Image of a vase

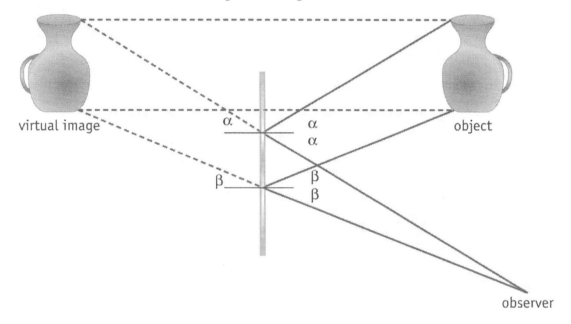

Figure 159 - Drawing the lines to locate the virtual image

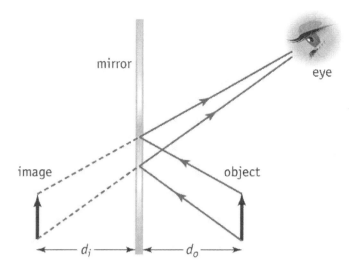

Figure 160 - the distances of the image and real object to the mirror are the same

Notice that in the right hand diagram above, the distance to the virtual image from the mirror is exactly the same distance away from the mirror as the original, and it is the same size as the original.

25.4 Questions

1) An object I is placed near a plane mirror. Draw this to the full size with two rays from O to the mirror. Draw in the two reflected rays and the position of the virtual image

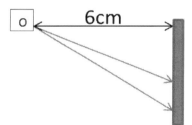

2) Write "AMBULANCE" laterally inverted
3) The diagram shows 2 rays of light shining into a mirror from a torch, T.

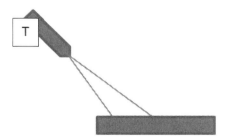

   a) Draw the two reflected rays
   b) Draw the image of the torch in the correct position
4) Name 3 properties of an image in a plane mirror
5) Name the two laws of refraction

6) Name the 2 types of reflection and give examples of each type

25.5 Refraction

- When light bends as a result of changing speed while entering a slower (or faster) media.
- The medium is usually, for light, a transparent material, such as glass, water, air or oil

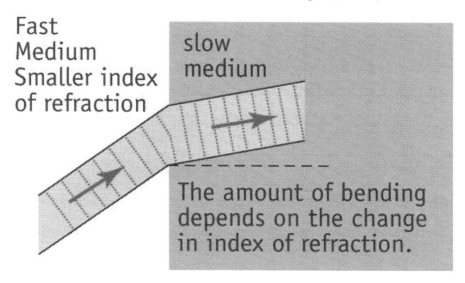

Figure 161 - refraction from fast to slow media

Typical experiments that explore this phenomenon use a small rectangular block of glass and a ray box. The ray box is then shone into the block and the light rays are traced on a sheet of paper placed under the equipment.

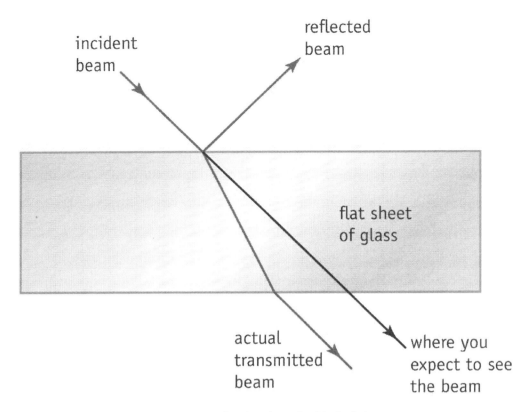

Figure 162 - refraction through a block of glass

The incident beam both refracts as it intercepts the glass as well as reflecting. The reflection is because of imperfections in the surface of the glass that cause a diffuse reflection (see above). Of immediate interest is the ray as it passes through the block and slows down. As it slows down, it changes wavelength and its direction. It obeys a rule know as Snell's Law. As the light enters the glass and slows down it bends TOWARDS the normal (see Figure 163 - Refraction in a solid block of glass) e.g. from air to glass. Its angle of incidence and angle of refraction are related through Snell's law and a quantity that relates how much the speed of light changes by. This quantity is called the refractive index.

Note how, when the light reappears and returns to its original speed, it bends AWAY from the normal. This is true for light as it moves from a slower media to a faster one, e.g. glass to air.

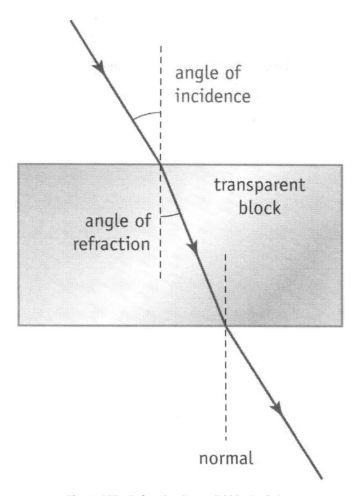

Figure 163 - Refraction in a solid block of glass

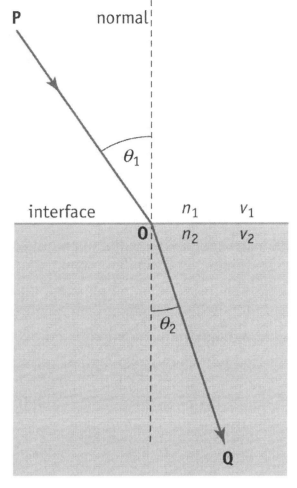

Figure 164 - exploring the angles in refraction

In this diagram we can see the two key angles, θ_1 and θ_2, are the incident and refractive angles respectively. The two speeds of light are represented with v_1 and v_2. In this diagram the incident ray is in a faster media than the refracted ray.

The refractive index, n, is defined as the ratio of the two speeds:

	Quantity	Name	Units
$$n = \frac{v_1}{v_2}$$	n	Refractive index	No units
	v_1	Velocity in the initial media	m/s
	v_2	Velocity in the second media	m/s

Snell's law also shows that the refractive index is

	Quantity	Name	Units
$$n = \frac{\sin \Theta_1}{\sin \Theta_2}$$	n	Refractive index	No units
	Θ_1	Angle of incidence	degrees
	Θ_2	Angle of refraction	degrees

Different materials have different refractive indexes, as seen in the table below:

Table 7- some typical refractive indexes

Medium	Refractive Index	Light Speed (m/s)
Diamond	2.42	1.02×10^8
Glass	1.5	2.00×10^8
Water	1.33	2.25×10^8
Perspex	1.49	2.01×10^8
Air	1.0003	3.00×10^8

The table above shows how the refractive index varies between materials, with diamond being the one that refracts most and where the speed is the lowest.

25.6 Calculating refraction indexes for a single substance

Snell's law shows that the ratio of the sines of the incident and refractive angles will be constant for a particular media. For instance, in the case of glass, the following table shows how the angles might vary, but the refractive index stays the same

Table 8 - Refractive angles for glass with n=1.5

Incident Angle (°)	Refractive Angle (°)	Sine of incident angle (i)	Sine of refractive angle (r)	Sin i/Sin r
15	10	0.26	0.17	1.5
45	28	0.71	0.47	1.5
27	18	0.45	0.30	1.5

Worked example

A light ray enters a transparent material at an angle of 34°, its angle of refraction is 22°, calculate the substance's refractive index, n.

$$n = \frac{\sin \Theta_1}{\sin \Theta_2} = \frac{\sin 34}{\sin 22} = \frac{0.56}{0.37} = 1.51$$

This shows that the media is most likely glass, as per table Table 7- some typical refractive indexes

Worked example

Light in air strikes water at an angle of incidence of 45°. If the refractive index of water is 1.33, calculate the angle of refraction.

Use Snell's law:

$$n = \frac{\sin \Theta_1}{\sin \Theta_2}$$

Rearranging the equation to find the angle, Θ_2, we get:

$$sin^{-1}(\frac{\sin \Theta_1}{n}) = \Theta_2$$

Putting the values into the resultant equation we get:

$$sin^{-1}(\frac{\sin 45}{1.33}) = 32°$$

25.7 Questions

1) The refractive index of water is 1.33. Calculate the angle of refraction of light that strikes at an incident angle of (a) 24° and (b) 53°
2) Examine the following ray diagram:

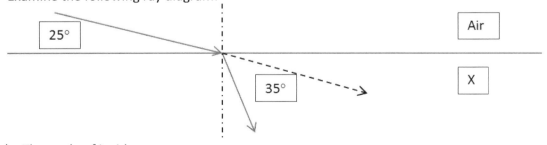

a) The angle of incidence
b) The angle of refraction

c) What is the refractive index of substance X

3) The table shows some angles of incidence and refraction for a substance.

Θ_1	0	10	20	30	40	50	60	70	80	90
Θ_2	0	7	13	19	25	30	35	39	41	42

4) Plot a graph of Θ_1 against Θ_2
   a) What angle of incidence would give an angle of refraction of 10°?
   b) What angle of refraction would give an angle of incidence of 36°?
   c) What is the refractive index of the media?
   d) What is the media (reference Table 7- some typical refractive indexes)?

5) Copy and complete the diagram below

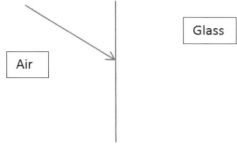

   a) Draw in and label the normal, the refractive ray, the angle of incidence and the angle of refraction
   b) How would the diagram be different if the rays were passing into water?

6) A light ray strikes a Perspex block at an angle of 60°; the refractive index of Perspex is 1.49.
   a) calculate the angle of refraction of the ray
   b) calculate the refracted angle for an incident angle of 39°
   c) what would you expect to happen if the incident ray struck the block along the normal?

Chapter 26. Total Internal Reflection (TIR)

This is a special case when dealing with refraction between a denser and less dense media, such as glass and air. When light moves from a slower to a faster media, there comes and angle where the light is no longer able to refract. This angle is called the **critical angle**. This angle is when Θ_2 is at 90° for an incident ray at incident angle, Θ_1. At this point all the light is reflected back from the denser substance.

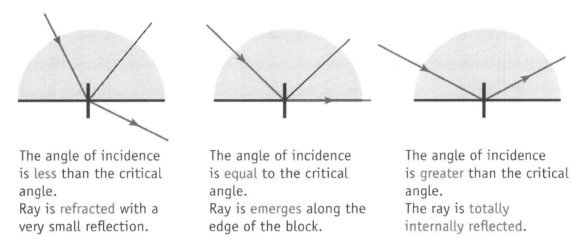

The angle of incidence is less than the critical angle.
Ray is refracted with a very small reflection.

The angle of incidence is equal to the critical angle.
Ray is emerges along the edge of the block.

The angle of incidence is greater than the critical angle.
The ray is totally internally reflected.

Figure 165 - Critical angle in a semi-circular block

Semi-circular blocks are often used to demonstrate total internal reflection because they provide an easy way of ensuring that the ray that moves from air into the block is always at 90° to the glass and will therefore not refract on the way in. This is achieved by aiming the ray at the centre of the flat surface through any part of the curved edge. These blocks are also easy to use to measure angles, as they are very similar to a compass.

- When the refracted angle is larger than 90°
- When it emerges from a denser media to a less dense media
- It is related to refraction
- It starts to behave like a mirror

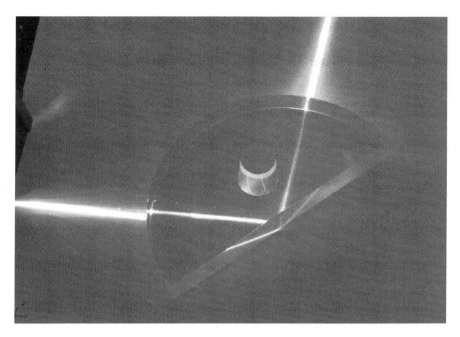

Figure 166 - Total Internal Reflection

26.1 Calculating the critical angle

Using Snell's law:

$$n = \frac{\sin \Theta_1}{\sin \Theta_2}$$

We need to consider carefully what is the incident angle and refractive angle when we use the usual refractive index. Usually the refractive index is larger than 1 and is the ratio of the speed of the faster media to the slower media. In this case we need to use it from the slower media to the faster media and so we use the inverse:

$$n = \frac{\sin \Theta_2}{\sin \Theta_1}$$

When the refractive angle is 90°, the incident angle is called the critical angle, Θ_c. And so, substituting in, we get Snell's law as:

$$n = \frac{\sin 90°}{\sin \Theta_c}$$

Since sin of 90 degrees is 1, rearranging this we get:

	Quantity	Name	Units
$\Theta_c = \sin^{-1}\left(\frac{1}{n}\right)$	n	Refractive index	No units
	Θ_c	Critical angle	degrees

Using the refractive indexes in "Table 7- some typical refractive indexes", we find that the critical angles for these media are as follows:

Table 9- refractive indexes of some common materials

Media type	Critical angle
Diamond	24°
Glass	42°
Water	49°
Perspex	42.1°
Air	85.5°

Figure 167 - - TIR in the surface of water - note the reflection of the tortoise in the surface

26.2 Uses of Total Internal Reflection

Using reflecting prisms

- The inside of a prism is used as a mirror when the incident ray is at more than 42°. This can be used for periscopes – these typically use an incident angle of 45°.

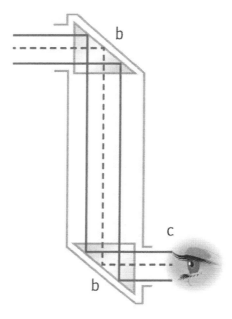

Figure 168 - Using two prisms that have 45 degree angles to the viewer allows someone to view an image higher up that is also the right way up

- Reflecting back using two 45° prisms, side by side, at opposite angles can create cats eyes of binoculars.

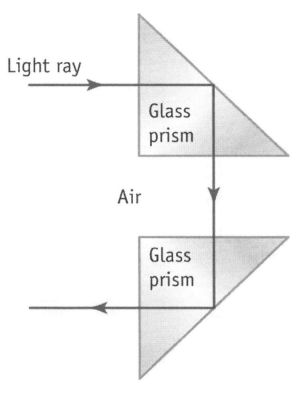

Figure 169 - How cat's eyes work using TIR

The advantage of using prisms for periscopes is because they avoid the ghost images that can occur when using glass mirrors.

26.3 Optical Fibres

- Uses a glass (or Perspex) fibre
- Light is shone into it, ensuring that the internal angles are greater than the critical angle
- Used in transmitting information and for viewing around corners.

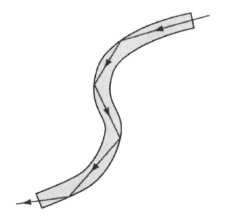

Figure 170 - total internal reflection in an optical fibre

Figure 172 - Optical fibres

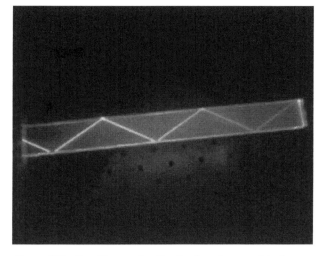

Figure 172 - Total internal reflection in a Perspex block

26.4 Questions

1) Glass has a critical angle of 42°, explain what this means
2) Copy and complete the following diagrams for glass blocks

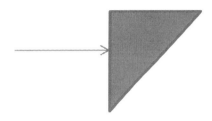

3) Give a practical example of the use of Total Internal Reflection
4) The diagram below shows a light signal travelling through an optical fibre made of glass:
   a) State 2 changes that happen to the light when it passes from the air into the glass at point B
   b) Explain why the light follows the path shown after hitting the wall of the fibre at P

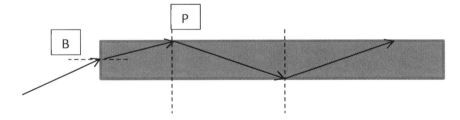

5) Copy and complete the diagrams below, for rays of light hitting the glass blocks.

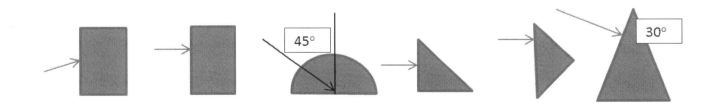

Chapter 27. Lenses

- Lenses bend light and form images as a result of varying the angle of refraction
- Two types of lenses

27.1 Convex Lenses

This is also known as a converging lens.

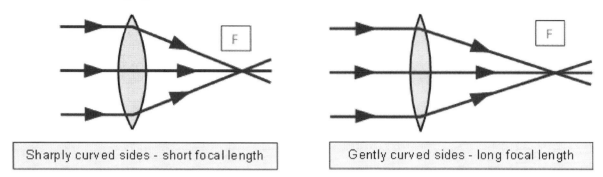

Figure 173 - Convex lenses

The convex lens has surfaces that bow outwards and the light rays converge at a point, called the **Principal Focus (F)**.

- Look through this at objects that are closer than the focal point will magnify them
- Looking at distant objects make them look upside-down

The distance between the centre of the lens and the principle focus is called the **Focal Length (f).**

27.2 Concave Lenses

This is also known as a diverging lens.

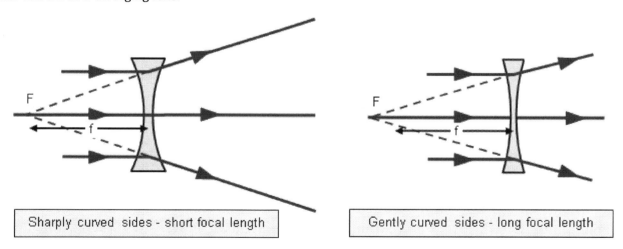

Figure 174 - Concave, or diverging, lens

The concave lens has surfaces that bend inwards, causing the light rays to spread out (diverge). The rays appear to come from the Principal Focus (F).

- The image you see will always be the right way up
- Virtual (cannot be formed on a screen)
- Smaller than the original

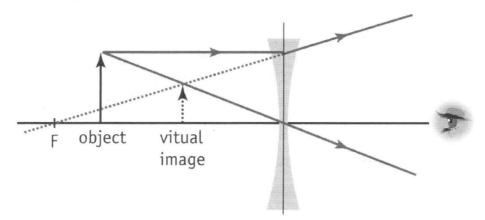

Figure 175 - Reduced image of larger object in a concave mirror

27.3 Real and virtual images

Real images are those where light actually converges, whereas virtual images are locations from where light appears to have converged. Real images occur when objects are placed outside the focal length of a converging lens or outside the focal length of a converging mirror

27.4 How images formed by lenses

- To predict where an image will be formed by a convex lens, us a ray diagram
- Only need to trace two of the rays
- Type and size of the image depends on where the object is paced in front of the lens

Distant objects – placed at a point >2F from the lens

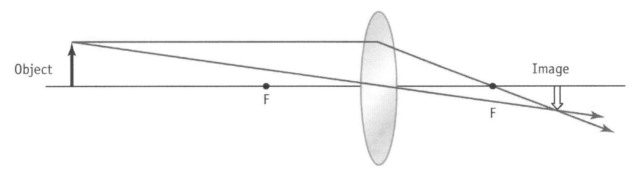

Figure 176 - converging lens where the object is beyond 2F

- Image is formed between F and 2F
- Image is
  - Inverted
  - Smaller than the original (Diminished)
  - Real
- Uses are in cameras

Object between F and 2F

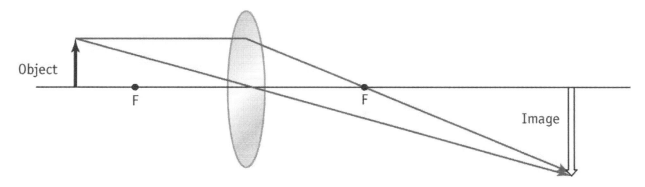

Figure 177 - coverging lens there the object is between F and 2F

- Images are formed beyond 2F
- Uses are in slide and film projectors
- Images are:
    - Real
    - Magnified
    - Inverted

Object between F and the lens

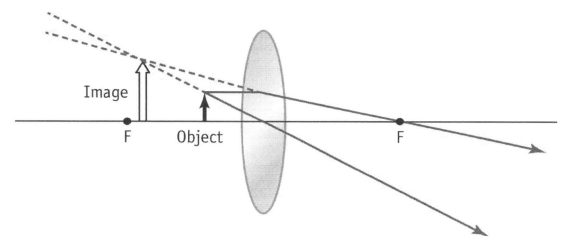

Figure 178 - converging lens where the object is closer than the focal point, F

- The rays never converge
- Uses in magnifying glasses and eye lenses in telescopes
- Image is:
  - Virtual
  - Upright
  - Magnified

Object location	Type of image	Orientation	Size
Beyond 2F	Real	Inverted	Diminished
Between F and 2F	Real	Inverted	Magnified

| Closer than F | Virtual | Upright | Magnified |

27.5 Dioptre – Lens Power

The distance to the focal point, f, is measured in metres. Lens makers tend to consider the "power" of the lens instead of the focal length.

$$dioptre = 1/f$$

Quantity	Name	Units
Dioptre	Lens power	1/m
f	Focal length	m

When manufacturing lenses, the lens maker will also describe the type of lens by either a positive value for a converging lens, or a negative value for a diverging lens, e.g:

+4.0D is a convering lens with a focal point at 0.25m (1/4.0)
-2.0D is a diverging lens with a focal point at 0.50m (1/2.0)

27.6 Worked ray tracing example

An object 2cm high stands on the principal axis at a distance of 9cm from a convex lens. If the Focal Length of the lens is 6cm, what is the Image's:

1) Position
2) Height
3) Type

First select a scale. For this plot 1cm=2cm, then draw the lines, with a ruler, in this order:

A – draw to the centre of the lens right along the vertical axis
B – draw this, through the centre of the lens and beyond it for distance
C – draw this through the Focal Point and until it meets line B
D – draw and measure this to the principal axis
Now measure the height of D and convert back using the scale

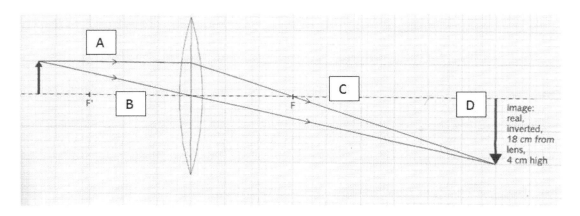

1) Answer is 18cm (9cm on the paper)
2) It is 4cm high (2cm on the paper)
3) It is real as the object is place beyond F but closer than 2F from the lens

27.7 Questions

1) Explain what is meant by:
   a) Principal Focus of a convex lens
   b) Focal length of a convex lens
2) If a convex lens picks up rays from a distant object,
   a) where is the image formed?
   b) If the object is moved towards the lens, what happens to the position and the size of the image?
3) An object 4cm high is placed 15cm from a convex lens with F=5cm. Draw a ray diagram on graph paper to find the position, size and nature of the image.
4) An object 3cm high is placed 3cm in front of a convex lens of F=6cm. Use a full scale to find the position, size and nature of the image.

Chapter 28. Dispersion of Light

- White light consists of the whole spectrum of colours
- A prism can be used to split them up into different colours
  - The reason this occurs is because different wavelengths of light have different refractive indexes. This is not covered in any GCSE or iGCSE syllabus.
- The range of colours is called a spectrum
- A glass prism is used to split the light apart
- The splitting is called **"dispersion"**

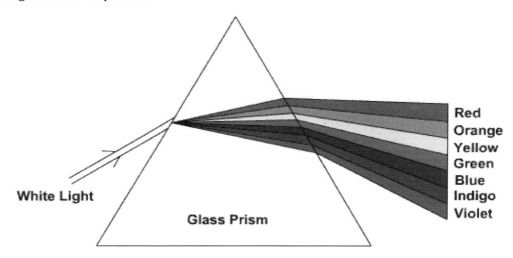

Figure 179 - dispersion (through refraction) in a prism

Up until this point we have assumed that the light that has been refracted has been **"monochrome"**, in dispersion we are dealing with light that contains multiple wavelengths, or colours. What we have not discussed to present is that different wavelength refract at different angles. The visible spectrum consists of 7 colours, memorised as ROYGBIV (red, orange, yellow, green, blue, indigo and violet). You could use "Richard of York grown bananas in vases" as a way of remembering this.

The longest wavelength refracts least, i.e. Red. The shortest wavelength refracts most. This occurs because the refractive index for each colour is different:

Table 10 - Refractive indexes for different colours

Colour	Wavelength	Index of Refraction in Glass
Blue	434 nm	1.528
Yellow	550 nm	1.517
Red	700 nm	1.510

Chapter 29. The Human eye

The human eye works by focusing light that enters it through the cornea at the front eye, through the eye lens, onto the retina, at the back of the eye.

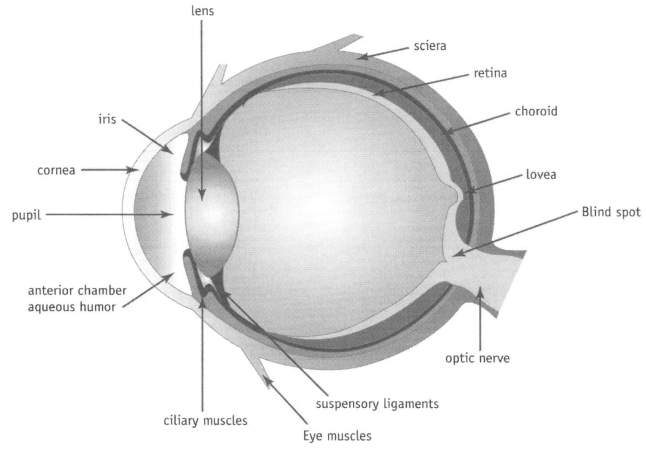

Figure 180 - diagram of the human eye

29.1 Parts of the eye

As can be seen above, the eye is built around aiming the light that enters it onto the retina. This is achieved by:

Cornea
Light enters the eye, initially, through the cornea. This partly focuses the light, but also protects the rest of the eye from damage.

Pupil
The pupil widens and tightens to allow enough light in to the eye.

Iris
The iris is a coloured ring of muscle that controls the width of the pupil and this is the controller for the amount of light that enters the eye.

The Eye lens
The lens is a convex, transparent, organic lens that can change shape so that light will be focused on the retina at the back of the eye

Ciliary muscles and suspensory ligaments
The eye lens is connected to the ciliary muscles by the suspensory ligaments. This allows the ciliary muscles to change the shape of the eye lens by stretching it, or causing it to contract. In doing so, the eye lens thickness changes, and so the rays from different distances are focused on the retina.

Retina
Light sensitive cells are located on the retina and the signals from these are sent to the brain out via the optic nerve. At the location the optic nerve connects to the eye there are no light sensitive cells and so none of the light that lands there is detected. It becomes a "blind spot".

29.2 Forming the image

The image seen at the back of the eye is upside down, but the human brain corrects for that. In the following image, you can see how the flower is turned upside down and then ends up being shown on retina laterally inverted.

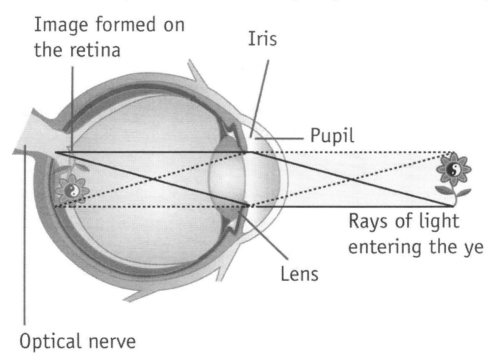

Figure 181 - Images at the back of the eye are laterally inverted

If the eye is mis-shapen, such as too flat, or stretched, the image will not land in the right place.

Short-sightedness (myopia)
When the image is formed in front, not on, the retina, because the eye is too long, and images that are far away become blurry.

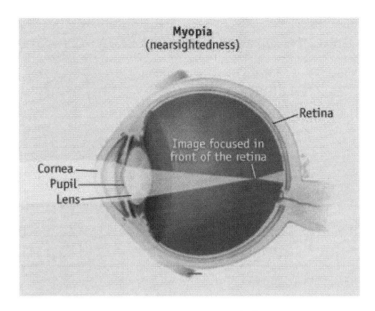

Figure 182 - myopic eyeball

Long sightedness (hyperopia)
When the image is formed behind the retina, because the eye has been flattened, images that are nearby are blurry.

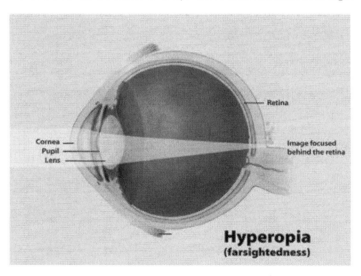

Figure 183 - Hyperopic eyeball

Correcting errors in the shape of the eye
Lenses can be used to correct the location of the focus of the image. The following image shows how this is done using standard lenses that we have learned about.

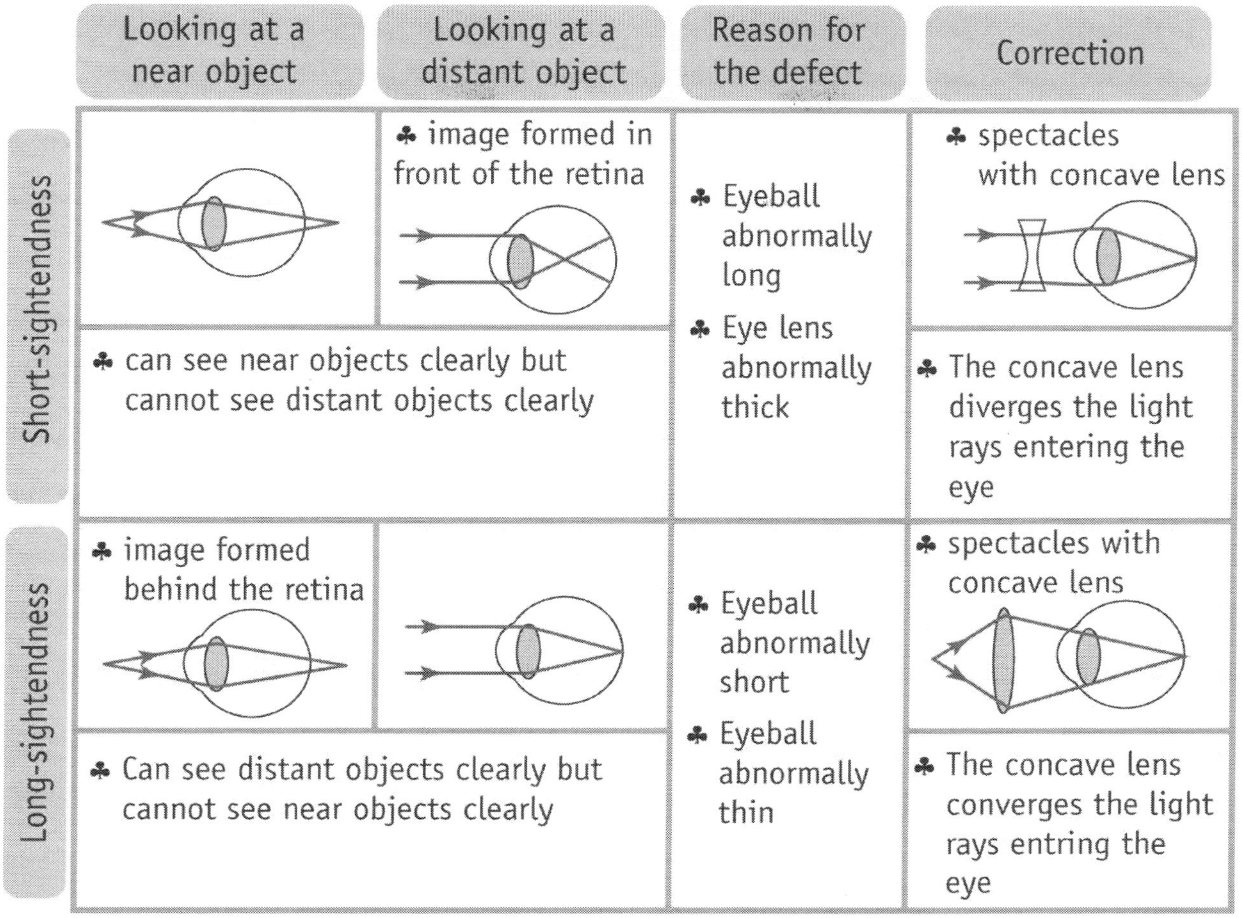

	Looking at a near object	Looking at a distant object	Reason for the defect	Correction
Short-sightedness		♣ image formed in front of the retina	♣ Eyeball abnormally long ♣ Eye lens abnormally thick	♣ spectacles with concave lens
	♣ can see near objects clearly but cannot see distant objects clearly			♣ The concave lens diverges the light rays entering the eye
Long-sightedness	♣ image formed behind the retina		♣ Eyeball abnormally short ♣ Eyeball abnormally thin	♣ spectacles with concave lens
	♣ Can see distant objects clearly but cannot see near objects clearly			♣ The concave lens converges the light rays entring the eye

Figure 184 - lenses used to correct short- and long sightedness

29.3 Comparing the eye with a digital camera

1) The camera uses a variable "aperture" to stop the light from entering, the eye uses the iris to control the amount of light that passes into the eye
2) Both have real, inverted, diminished images
3) The camera uses a fixed focus converging lens (sometimes in pairs), the eye has a variable focus convergin lens
4) The camera uses a manual (or computerised) adjustment of the lens position to adjust the focus, the eye alters the thickness of the lens, using the ciliary muscle
5) The camera either has a CCD sensor (for a digital camera) or a photographic film to detect and store the image, whereas the eye uses light sensitive cells on the retina.

	The camera	The eye
Controlling the brightness	An aperture is opened more or less	The iris adjust the opening into the eye
The image	Both have real (the rays converge), upside-down and diminished images	
Type of lens used	Fixed focus convex lens	Variable focus convex lens
Focus adjustment	Movement of the lens position by either a computer or by hand	Thickness of the eye lens adjusted by the ciliary muscle
How image is caught	Using either a photographic film, or by CCD chips that record the image to memory	Light sensitive cells in the back of they eye, on the retina

Chapter 30. The Electromagnetic Spectrum

Light is but a small part of larger, invisible, spectrum, called the Electromagnetic spectrum.

This spectrum shares these following characteristics with light:

1) Can travel through a vacuum
2) Is a transverse wave
3) Travels at c, the speed of light (3×10^8 m/s)
4) Transfers energy

The electromagnetic spectrum is usually split into 6 different parts for ease of memory. The boundaries between these regions are agreed by international scientists. The diagram below shows how the wavelength varies between each region and the order that the spectrum appears in. It also shows how the shorter wavelengths (higher frequency) have increasing energy.

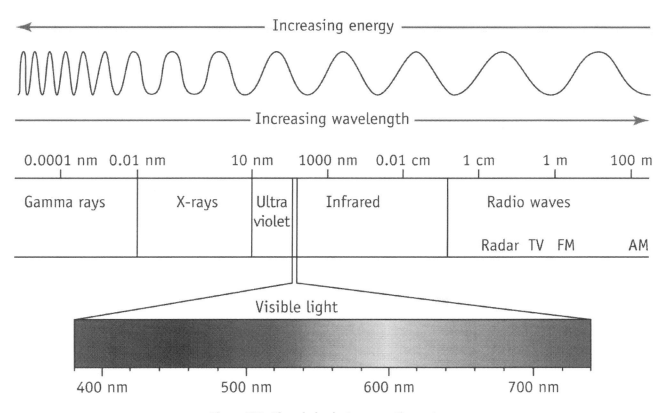

Figure 185 - The whole electromagnetic spectrum

The tables below show the agreed wavelength boundaries and the most common uses of that region of the spectrum.

Table 11 -Common uses for types of radiation

Wave	Wavelength	Use
Radio waves	>30 cm	TV transmissions, Radio and communication
Microwaves	3 cm	Communication

Infrared	3 μm	Radar Heating up food Communication in optical fibres Remote Controllers Heating
Light	400 - 700 nm	Seeing, Communicating
Ultra violet	100 nm	Sterilising, lamps, Forgery detection Sun tanning
X-ray	5 nm	Shadow pictures of bones
Gamma rays	<0.01 nm	Scientific research, kills cancer cells, sterilising medical equipment

Table 12 - Types and levels of hazards from exposure to the different categories of EM radiation

Wave	Hazard	Prevention
Radio waves	No hazard	
Microwaves	Heating of water in the body	Metal grid
Infrared	Heating effect	Reflective surface
Light	No hazard, LASERS can cause blindness	Clothes
Ultra violet	Can cause cancer	Sun cream (or cover up)
X-ray	Causes cell damage	Lead screens
Gamma rays	Causes cell damage	Thick lead screens or concrete

Gamma Rays
- Come from radioactive sources, e.g. Uranium, historic nuclear bomb testing and from stars
- Highly penetrative – only lead can protect, but not completely

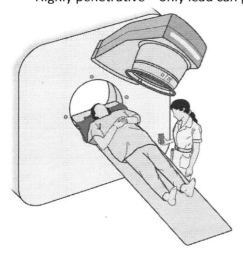

Figure 186 - Radiotherapy uses gamma ray to treat cancer

X-Rays
- Produced by bombarding certain metals with high energy electrons

- The main difference between X-rays and Gamma rays is how they are produced
- Highly penetrative

Ultraviolet
- Produced on the Sun – causes tanning, in mercury lamp and from sparks
- Low penetration

Visible light
- Created by lamps, stars, heaters...
- Spectrum of colours
- No penetrative effects

Infrared
Created by electric fires, sun, TV remotes

Microwaves
- Created by microwave ovens, stars, TV and communication satellites

TV and Radio
- Sources include stars, some planets such as Jupiter, TVs, Radios and mobile phones.

Questions

1) List 3 properties common to all electromagnetic waves
2) Place the following in order of wavelength, the longest first: Ultraviolet, X-Rays, Violet Light, Gamma rays, Radio, Infrared, Microwaves
3) Define dispersion
4) Why are different colours produced through a prism
5) The diagram below shows the main areas of the Electromagnetic spectrum:

Radiowaves	Micro waves	B	VL	UV	A	Gamma rays

a) Name Regions A and B
b) Write down, in words, the equation connection wave speed, wavelength and frequency
c) Calculated the frequency of the radiation with wavelength of 0.001m given that the speed of all EM waves is 3×10^8 m/s.
6) Explain how and why gamma rays can cause damage to living cells
7) Use the following words in the sentences below:
[LESS THAN], [THE SAME AS] and [GREATER THAN]
The wavelength of radio waves is _____ the wavelength of Ultraviolet radiation
In a vacuum, the speed of ultraviolet radiation is _____ the speed of light
The frequency of gamma radiation is _____ the frequency of infrared radiation
8) Name the spectrum of colours (least first) in order of deviation
9) "Radio waves have the highest frequency and shortest wavelength" – TRUE/FALSE?

Chapter 31. Sound waves

- Caused by vibrations – e.g. loud speakers, tuning forks, stringed instruments, voice box
- Sound waves are longitudinal waves
- Vibrations are parallel to the direction of movement

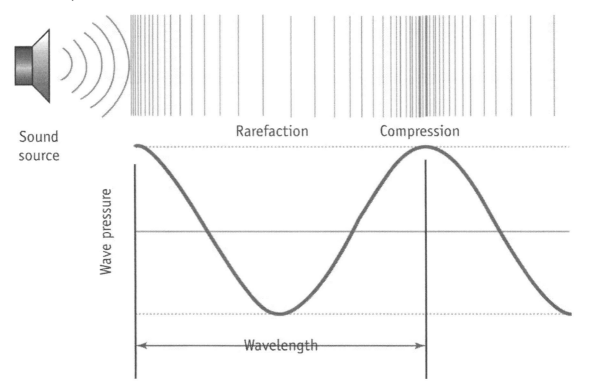

Figure 187 - Sounds waves - showing rarefaction and compressions.

Figure 188 - Shock waves, such as these from the guns of a battle ship, are single, very highly compressed areas of air particles that can be damaging to structures and people.

- Areas where the particles in the wave are squashed together (high pressure areas) are known as compressions.

- Areas in between the compressions are called rarefactions and the pressure there is low
- Sound cannot travel through a vacuum
- Sound waves can travel through solids, liquids and gases

Experiment to show that air molecules are required to transmit sound waves
Use a vacuum pump

- All air is pumped out of the bell jar
- A ringing bell ceases to be heard

As the air is evacuated you will gradually stop hearing the ringing, as air is let back in the sound will yet again be audible.

Speed of sound and echoes
- In air, at 0°C, the speed of sound is 330m/s
- It depends on:
  - Temperature of the air
  - Does not depend on the pressure
  - Is different in different materials – fastest in solids, over 1000m/s

Measuring speed of sound

The formula to use is:

$$v = \frac{distance\ (m)}{time\ (s)}$$

Example: A sound wave travels 1500m in 10 seconds, calculate the speed:

$$v = \frac{distance\ (m)}{time\ (s)} = \frac{1500}{10} = 150m/s$$

Reflection of sound and echoes
- Hard surfaces can reflect sound waves
- If the reflected sound is heard a short time after the original sound, and echo has been heard
- In an echo, the sound wave has to travel to the surface and back again (time taken is therefore twice the time required to reach the reflecting surface)
- The distance to the object can be measure if the speed of sound and the echo time is known
- Used in echo-sounding, e.g. depth of oceans for a surface boat, or locating obstacles on a submarine

31.2 Worked examples

1) A man claps his hand near a wall. He hears his clap echo back after 2s. Calculate the distance to the wall if the speed of sound is 330m/s.

$$velocity\ x\ time = total\ distance = distance\ to\ wall\ x\ 2$$

$$distance = \frac{velocity\ x\ time}{2} = \frac{330\ x\ 2}{2} = 330m$$

2) A ship sends out a sound wave to the ocean floor and hears an echo 7s later. How far is the ocean floor if the speed of sound in water is 1500m/s?

$$velocity \times time = total\ distance = distance\ to\ ocean\ floor \times 2$$

$$distance = \frac{velocity \times time}{2} = \frac{1500 \times 7}{2} = 5250m$$

3) A submarine sends a sound wave into the water and echo is returned 0.2s later. The speed of sound in water is 1500m/s. Calculate the distance to the obstacle.

$$velocity \times time = total\ distance = distance\ to\ obstacle \times 2$$

$$distance = \frac{velocity \times time}{2} = \frac{1500 \times 0.2}{2} = 150m$$

31.3 Terminology of sound waves

1. Frequency – number of waves per second (s^{-1} or Hz)
2. Pitch – how high or low as sound is – related to the frequency
3. Amplitude – the vertical size of the wave from the equilibrium position (rest position)
4. Loudness – When the amplitude changes, the sound is more or less loud

Humans can hear sounds in the range of 20-20000Hz

31.4 Oscilloscopes tracing sound waves

The form of the wave on an oscilloscope screen is difference depending on its wavelength, amplitude and frequency.

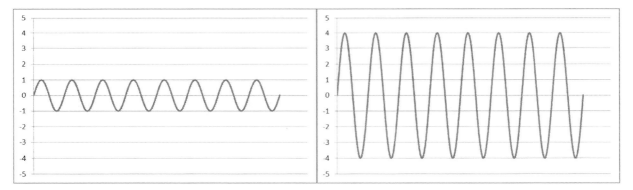

Figure 189 - High Pitch, quietFigure 190 - High pitch, loud

The two Oscilloscope traces above shows a high frequency (pitch), with different amplitudes. This means that one oscilloscope is tracing the sound of a loud, high pitch sound and the other is tracing the same pitch, but quieter.

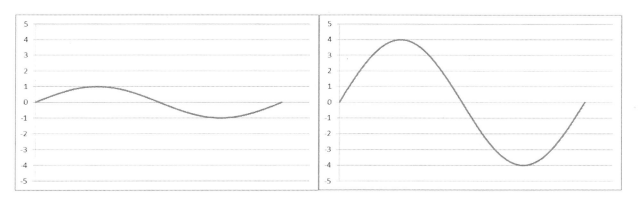

Figure 191 - Low pitch, quiet   Figure 192 - low pitch, loud

31.5 Questions

1. What is the wavelength of a sound wave of frequency 100Hz? (speed of sound – 340m/s)
2. Define amplitude, wavelength and frequency
3. Draw an oscilloscope trace for a loud, high sound
4. Describe how the sound from a vibration tuning fork travels to the ear of listener
5. A burst of sound waves is emitted from the fishing boat. The echo from the shoal is detected 0.4s later. Sound in water travels at 1500m/s
   How far has the pulse travelled in 0.4s
   Calculate the depth of the shoal
6. Below are the frequencies of 4 sounds:
   1. 400Hz   2. 150Hz   3. 500Hz   4. 200Hz
      a. Which sound has the highest pitch
      b. which sound has the longest wavelength in 0°C air?
7. Draw an oscilloscope trace for a quiet, low frequency sound
8. Distinguish between longitudinal and transverse waves
9. Calculate the frequency of a sound wave with wavelength 2cm, travelling at a speed of 340m/s
10. A rod 2.4m long is being struck with a hammer at one of its ends. Four measurements of the time interval between striking it and the sound arriving at the other end were 0.44ms, 0.50ms, 0.52ms and 0.47ms:
    a. Determine the average value of the four measurements
    b. Calculate the speed of sound in the rod

Chapter 32. The Atom and the Electron

32.1 Atomic Structure

The atom will be explored in more detail, but to help understand how electricity and magnetism works, it is important to understand something about atoms, and, especially, electrons. An atom has 2 major parts:

- The nucleus – which consists of varying numbers of neutrons and protons. The nucleus has an overall POSITIVE charge
- The Shells – these are electrons that orbit, or spin, around the nucleus in different shells.
    - The outer shell is called the valence shell and it is the electrons in this shell that is involved in electricity, magnetism, the transfer of heat and in most chemical bonds.
    - The shell consists of NEGATIVE particles; electrons.

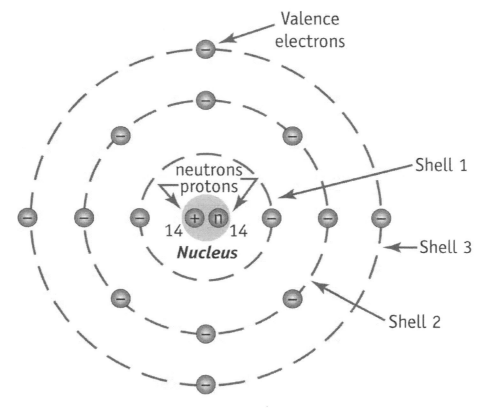

Figure 193 - The model of the atom

The protons and the neutrons are assumed to vibrate together with their shells *in situ*, when they are part of a solid object, such as a metal. You will be shown how, sometimes, a nucleus can end up separate from its electrons, but in for now, and in electricity and magnetism, the nucleus is ALWAYS in the same place – it does not move (except for the vibration *in situ*). When we explore electricity, remember, the nucleus, and therefore the protons DO NOT move. Valence electrons (those in the outer shell), however, are only loosely attached to the nucleus and can be released to allow them to move away.

Metals, as you will find out in chemistry during the metallic bonding topic, have "free" electrons that can easily move between the **lattice** of the metallic element.

32.2 Electrons

These wonderful particles are involved in so many parts of our everyday life and were only relatively recently discovered, although several years before the proton and neutron. They have various properties:

1) They are charged – they have a negative charge of the same size as the proton's positive charge.
2) They are very small, in fact, they might not have a size at all
3) They spin around themselves and around the nucleus
4) The repel each other because they have the same charge

Chapter 33. Magnets and Magnetism

33.1 How the electron is involved

When the electron spins around its own axis, its moving charge forms a very small magnet. When many of these electrons spin in agreement, they form a larger single, but very small magnet, called a **domain**. If enough of these domains end up lining up in the same direction, a larger magnet is formed.

In the diagram, the arrows describe how individual electrons spin in the same direction, but do not become a fully formed magnet until all the domains line up (as at the right).

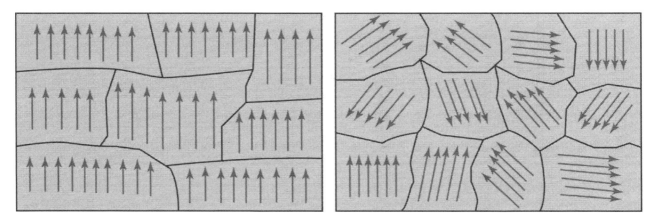

Figure 194 - How to visualise magnetic domains and their magnetic orientation

33.2 Properties of magnets

Most magnets that you will have seen have two ends coloured in with red and white paint. These are the opposite **poles**, which are called NORTH and SOUTH. These poles either attract the opposite pole, e.g. north attracts the south pole, or they repel each other, e.g. the south pole repels another south pole. The rule is:

- Like poles repel, unlike poles attract.

Only certain elements are able to form magnets, these are iron (Fe), nickel (Ni) and cobalt (Co). These elements have a special combination of electrons in the valence shell, where the electrons spin in agreement. In addition, Steel, which has other elements mixed with iron, can also form a magnet. These materials are called **ferromagnetic**.

- Iron, Nickel, Cobalt and Steel are magnetic materials.

The magnetic poles exert forces on other magnetic objects at a distance due to a **magnetic field**. This field can be visualised with arrows that go from the North Pole to the South Pole. These lines of force are also called lines of **flux**.

- A magnetic field are force lines, flux lines (B), with a direction – North to South

Magnets can be used to **induce** magnetism in other objects by carefully rubbing a magnet on a non-magnetic object.

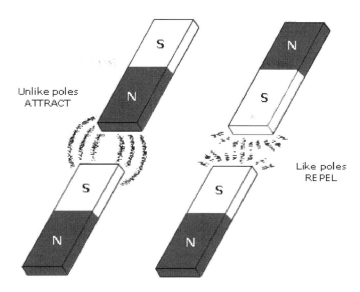

Figure 195 - How permanent magnets affect each other

Magnetic flux lines

The word seems scary, but it simply is a word to describe the existence of lines of force, which are shown are lines, with arrows going from the North Pole to the South Pole of a magnet. The stronger the magnet, the more closely these flux lines can be considered to be:

- The density of the flux lines shows how strong a magnetic force is in existence at that point

Ferromagnetics

These types of magnets are either **hard** or **soft** depending on how easily they LOSE their magnetism:

- Soft magnets are easy to make and just as easily lose their magnetism
    - **Iron** is an example of a soft magnetic material
    - Used in **transformers** and **electromagnets**
- Hard magnetic materials are difficult to magnetise and also difficult to demagnetise
    - **Steel** is an example of a hard magnetic material - although it contains iron, the other elements mixed in resists the domains from being able to realign
    - Used in compasses and permanent magnets

Hard and Soft Magnetic Materials

HARD	SOFT
E.g. Steel, alnico (aluminium-cobalt-nickel alloy)	E.g. Iron, mu-metal (a nickel-iron alloy)
Difficult to magnetise	Easy to magnetise
Keep their magnetism	Lose their magnetism
Used for permanent magnets, compasses	Used in transformers and electromagnets.

33.3 Magnetic induction – making magnets

Magnets are made by encouraging the domains to align, lining up in the same direction. There are two ways of doing this:

Stroking Method

As the permanent magnet is stroked over a non-magnetic material, the domains are gradually aligned with the stroking motion, due to the force of the magnetic field

- Steel or iron stroked with a bar magnet.
- North and south poles are INDUCED.

This is the oldest way of making magnets.

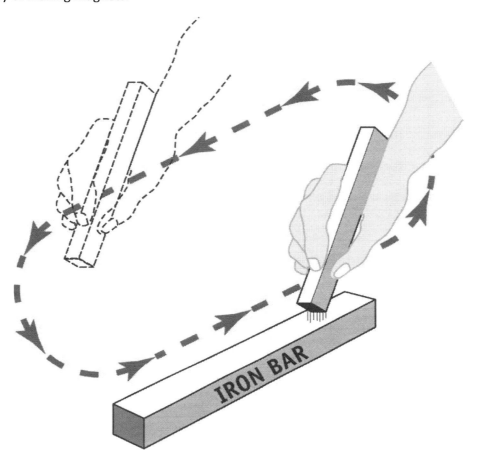

Figure 196 - Stroking a magnetic material with a permanent magnet will cause it to become a magnet

Electrical Method

When a current is passed through a coil wire, it provides a temporary, but powerful magnetic field around itself. This field will force any domains within it to line up with the magnetic field.

- Steel/iron placed in a coil of wire.
- D.C. current passed through.
- Current induces magnetism.

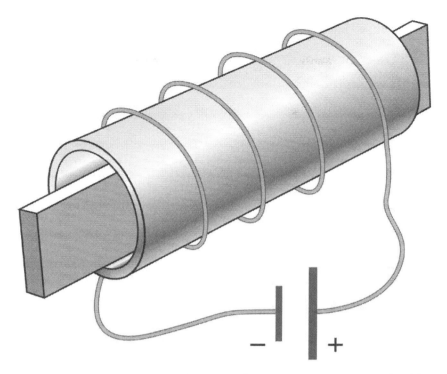

Figure 197 -A coil will produce a magnetic field that will make any magnetic material inside a magnet

33.4 Destroying magnetism – demagnetisation

All the methods of destroying a magnet involve causing the domains to, chaotically, mix up their alignments and thus stop their alignment with each other to cause a coherent magnetic field.

Hammering

When a magnet is hit by a hammer, the particles within it shake and can be caused to change their alignment. Hitting a hammer on a magnet several times will gradually cause the domains to realign and the material to lose its overall magnetic alignment. The same can happen if a magnet is dropped onto a hard floor.

Heating

Heat will cause the electrons to become more agitated, to move faster and more chaotically, and so the domains can end up losing their alignment.

Coil of wire using A.C. current

The current in a coil will cause a magnetic field in a particular direction when the current flows one way. If the current changes direction, it will cause a magnetic field to change direction too. Doing this often with a magnet inside the coil will cause the domains to become misaligned, thus losing its overall magnetism.

An A.C. current is a current that changes direction regularly.

33.5 Magnetic field

All magnets have an invisible force field around them that are called **magnetic fields**. When two magnets repel, or attract, each other, it is these force fields that interact. These force fields are illustrated as **lines** that start from the North Pole and end at the South Pole (and then back to the North Pole). These lines are drawn with **arrows** pointing from North to South, and are called **lines of flux**.

Observing magnetic field lines

You can see these lines (but not the direction) by using iron filings, sprinkled on a piece of paper, above a bar magnet. The magnetic iron filings each form a small magnet, lining up along the field lines, with their small North Pole pointing to the South Pole of the bar magnet. They connect to other iron filings as these are also small magnets, connecting north to south on a microscopic scale.

Observing the direction of the field lines

By adding small magnets that can rotate, known as plotting compasses, it is possible to determine which direction is south on a magnet, since the north pole of the rotating magnet will be attracted to it. Since lines go from north to south, the direction the rotating magnet points is the direction of the magnetic field.

Looking at the picture, you will notice that the lines converge at the poles, and the lines become very compressed. This increase in the density shows that the strength of the field is also strongest nearest the poles.

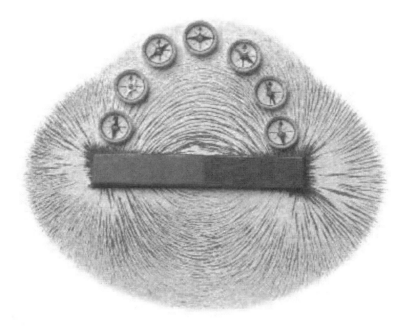

Figure 198 -tracing the magnetic flux lines using iron filings and plotting compasses

When magnets attract

The magnetic field lines from the North Pole join those entering into the South Pole of another magnet and become continuous magnetic field lines. This joining of field lines is a result of the attraction between the poles. They form a, continuous, single field, thus forming an attractive force.

Figure 199 - Magnets that attract. Note how the field lines join together

When magnets repel

The magnetic field lines are NOT ALLOWED to cross and since they are in opposite direction when like poles come close together, the lines will avoid those of the other magnet and deflect away from the central area between the two poles. This is because of the repelling force between like poles and explains why they repel.

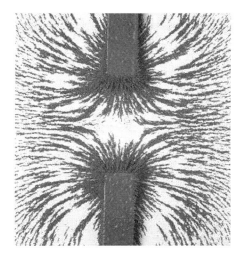

Figure 200 - Repelling magnets. Note how the field lines avoid each other

The area in the very centre, where the lines of force all avoid, is called the **neutral point**. Here the magnetic force is ZERO.

Figure 201 - The Earth's Geographic north pole is a magnetic south pole

The Earth's magnetic field

Since North Pole of a compass points towards the arctic on Earth, it follows that it is being attracted by the pole at the top of the world. And since opposites attract, it means that the Earth geographic North Pole is a magnetic South Pole, and the geographic south it a magnetic North Pole.

Plotting the magnetic field

To do this you will require a bar magnet and a small plotting compass.

1. Place the compass at the N-Pole.
2. Mark position of needle with 2 dots.
3. Move compass to needle lines up with previous dot.
4. Join the dots to produce a MAGENTIC FIELD LINE.

Further field lines can be drawn by starting the compass IN DIFFERENT POSITIONS.

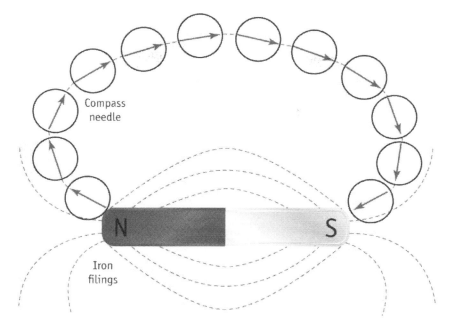

Figure 202 - Plotting the magnetic field around a magnet

33.6 Magnetic effect of a current

As mentioned at several places above, when charges move (known as a **current**), it causes (**induces**) a magnetic field to exist around it. In the following picture you can see a wire come straight up from underneath a Perspex plate, through a hole in it, and a current passed through it. Iron filings are then sprinkled around it. Note how they align in the same way as they did around a bar magnet.

- A magnetic field is present
- It circles around the wire
- The magnetic field is there going round in a continuous field, pointing North

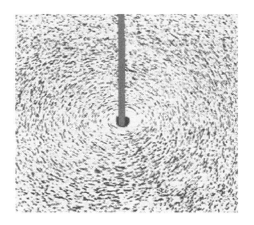

Figure 203- Note the concentric flux lines

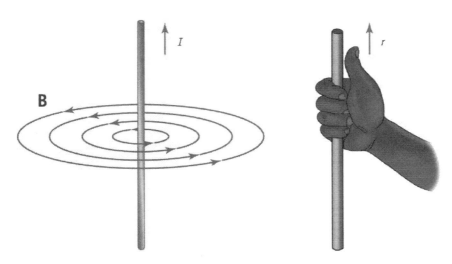

Figure 204 - Right hand thumb rule

By adding a plotting compass to this, you can find out the direction of that magnetic field. It is possible to work the direction of the current (I). This is called the **Right Hand Grip Rule**. By gripping a wire with the right hand, so that the fingers follow the field lines (North to South), and stretching the thumb along the wire, the direction of the current is the direction the thumb is pointing. You can also work out the direction of the magnetic field by gripping the wire, pointing the thumb up along the current. The direction the fingers curl around the wire is then the direction of the magnetic field.

The field strength is strongest close to the wire, and can be increased by increasing the current.

33.7 The coil

When a wire is wrapped around again and again, it is called a coil. If a current is passed through this wire, there will be a magnetic field induced, that has the same shape as a bar magnet.

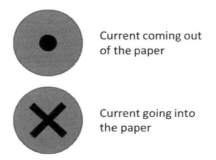

Current coming out of the paper

Current going into the paper

Before we can explore the reason for this we will introduce a way of illustrating a 3D concept – current moving into and out of the paper you are reading. Scientists have used the idea of looking at a dart from the front (the sharp tip towards the reader) as current coming OUT of the paper, and the dart from behind to illustrate the current going INTO the paper. Looking at the illustrations to the left, the upper circle shows how the current is coming out of the paper.

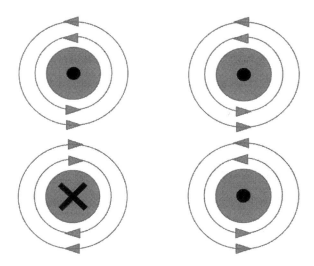

If we now add the magnetic fields to these two, we will see how the end up going in the same direction between them. The direction of the magnetic field is determined using the right hand grip rule (try it yourself). An interesting phenomena takes place when two wires are placed side by side with their magnetic field going in opposite directions. This, of course, as we saw with bar-magnets is not allowed and the field lines instead join into a continuous field.

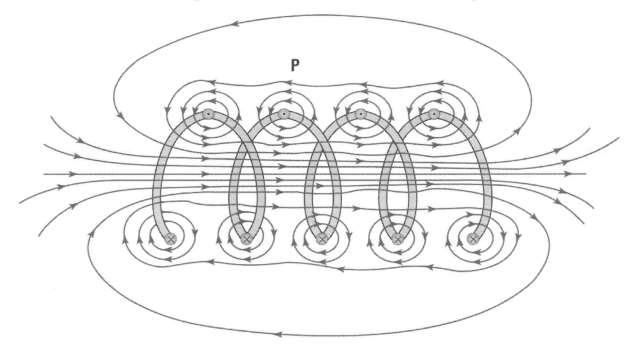

Figure 205 - Magnetic flux lines in a coil

33.8 Electromagnets

Since it is possible to create a magnet by switch on a current through a coil of wire, it is possible to create temporary magnets that can be used in many ways:

- Cranes to lift iron scrap
- Sorting machines to sort out magnetic materials from non-magnetic materials

- Doors that unlock when the power is off, such as fire exits
- Relay switches

To create an electromagnet you could just use a coil of wire, but it become a LOT stronger if you insert a soft iron core. You can increase the field strength by increasing the current or by increasing the number of turns in the coil. If you reverse the direction of the current through the coil, the magnetic field changes direction.

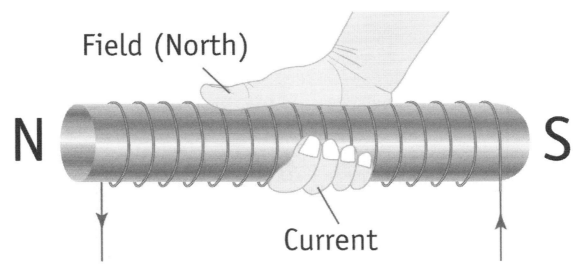

Figure 206 - The right hand coil clasp rule - shows the direction of an electromagnetic field produced by a coil of wire

There is a simple rule that is very similar to the right hand grip rule that helps you determine the direction of the magnetic field in a coil. Using the right hand coil clasp rule, grip the coil so that the fingers follow the current around the coil, and stretch out the thumb along the coil. The field is in the direction of the thumb.

Uses of electromagnets
Magnetic relays

Relays are switches using an electro-magnet to turn on the current in a more powerful circuit, such as a major piece of machinery. This is usually done to protect the user from high voltages.

- 1st Switch closed → current flows in the circuit.
- Coil magnetized → iron armature pulled close to the coil
- Iron armature acts as lever and closes the 2nd switch → larger current in main circuit.

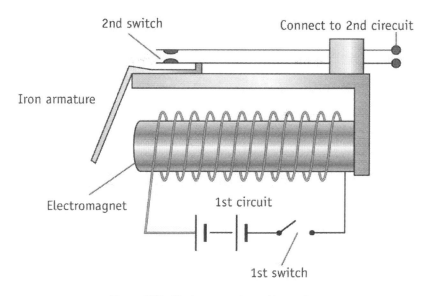

Figure 207 - Electromagnet used in a relay

Electric bells

This makes use of A 'Make And Break Circuit'.

1. Switch closed → current in electromagnet.
2. Hammer pulled across → strikes gong.
3. Contacts separate → electromagnet off.
4. Hammer springs back and close contacts
5. Repeating steps 2-4 until the switch is released

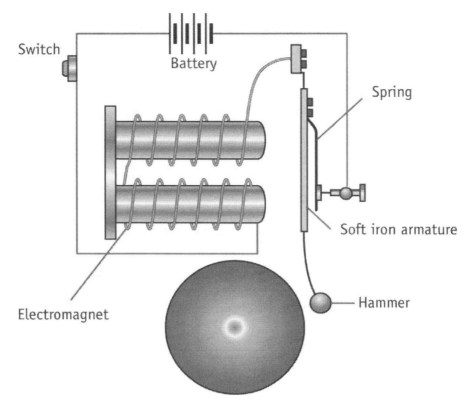

Figure 208 - Door bell using an electromagnet

Circuit breakers

This is used to cut off current in a circuit if too high. Used as safety devices in the home.

1. Current flows through the contact and the electromagnet
2. If too high a current, electromagnet's pull becomes high enough to release iron soft iron armature
3. Contact open, current stops
4. Contacts closed by pressing reset button.

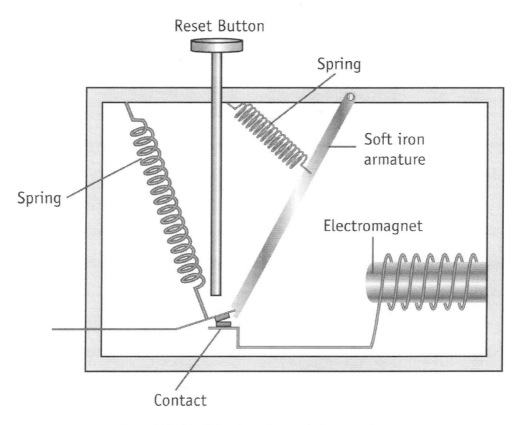

Figure 209 - Circuit breaker using an electromagnet

Chapter 34. Electrostatics

When there is a surplus (or lack) of electrons, a material is said to be **electrically charged**. If this charge is not flowing, it is said to be **static**, and so this charge is called **static electricity**.

34.1 Electrostatic materials

You will already have noticed that the structure of a material is important in determining how it will react to static electricity. E.g. metals are conductors since the electrons are loosely attracted to the nucleus, and are involved in weak metallic bonds and so can flow through the lattice. However, insulators have their electrons involved in much stronger bonds, e.g. covalent bonds (such as in plastics) or Ionic bonds (as in many crystals). Because the electrons are involved in strong bonds, they are very difficult to move and so these become insulators.

It is possible to add (or remove) electrons to **insulators**, but conductors will simply conduct the electrons, if possible, to another location.

When a **polythene** rod is rubbed with a cloth, electrons are transferred from the cloth to the rod, leaving it with a negative charge. The cloth is left with too few electrons, since they were transferred to the rod, and so becomes positively charged

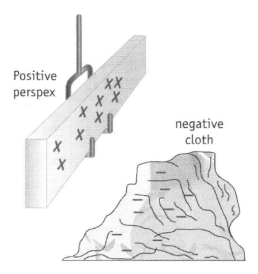

Figure 210 - Perspex and cloth

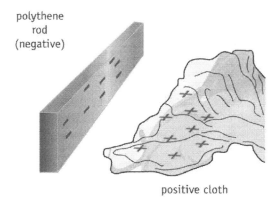

Figure 211 - Polythene rod with a cloth

However, when you do the same with a **Perspex** rod, electrons are transferred away from the rod into the cloth into the cloth. This leaves the rod with an overall positive charge, whereas the cloth now has a surplus of electrons, and so becomes negatively charged.

Attraction and repulsion

When the positive cloth, from the polythene rod rub above, is brought neat the rod, the rod will move **towards** the rod – they are **attracting** each other. However, when you bring two charged polythene rods, or two charged Perspex rods from above together, they will move away from each other – they are **repelling** each other.

- Like charges repels (e.g. negative repel negative)
- Opposite charges attract (positive charges attract negative charges, and vice versa).

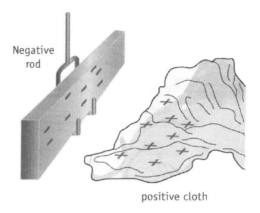

Figure 212 - Opposite charges attract each other

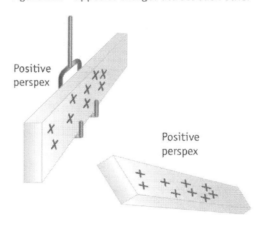

Figure 213 - Same charges repel each other

34.2 Conductors and insulators

Conductors are materials that allow electrons to flow through it with little effort, whereas insulators resist the flow of electrons. This is down to the configuration of the electrons in the bonds between the atoms.

Conductors	Insulators
Metals → Good such as copper, silver	Plastics, glass, rubber.
Have loosely held electrons.	Have tightly held electrons.
Electrons free-moving.	Electrons not free-moving.
Difficult to charge by rubbing.	Easy to charge by rubbing.

34.3 Electric Fields

Charged particles experience forces from each other. This is represented using field vector lines (arrows) in the direction a positive particle would move. Examples of these field lines are given below.

For point charges they are "radial" – they emanate from the centre

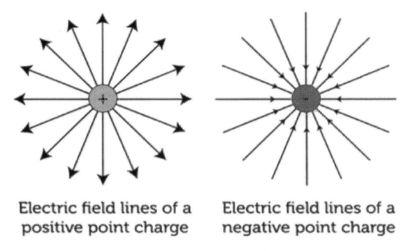

Figure 214 - Electric field lines

When two of these charges interact with each other, the fields interact. The law that must be followed is that field lines may not cross. If they conflict with each other, they repel, if they can connect, they attract.

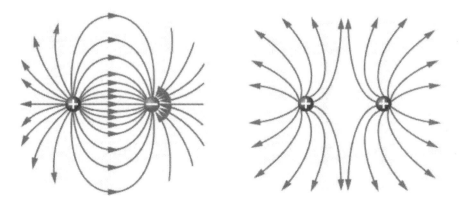

Figure 215 - Attraction and repulsion between point charges.

Between parallel plates, it is easier, and are simple parallel lines.

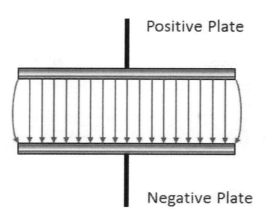

Figure 216 - Electric field lines between parallel plates

34.4 Attraction of uncharged objects

On occasions, when a charged object is brought near an uncharged object, the two of them will attract. This is because of a process called **charge induction**.

This can be seen in the example of a balloon and a charged rod, as shown below:

- The rod as an overall positive charge that attracts the electrons in the balloon.
- These electrons move, slightly, towards the surface at the end where the rod is
- The surface of the balloon is now negatively charged
- The rod and the balloon now attract each other

This charge on the balloon was **induced** by the rod being brought near it.

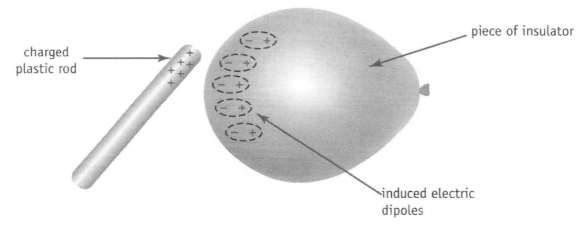

Figure 217 - Charge is induced on the surface of the balloon by the attraction of the electrons to the rod from the balloon

This also works for conductors, such as aluminium foil, as long as the aluminium foil is placed on an insulator so that it does not gain, or lose, electrons.

Earthing

Figure 218 - Lightning is caused by the earthing of charge from clouds to the Earth

When enough charge is built up on something, such as a balloon, or hair, the electrons will have enough repelling force between them, and enough attraction to another, lower charged object, to be pushed and pulled through air, or through another insulator. In air this will produce a spark.

This can be seen most dramatically in the charge that is built up in clouds, and the charge being released in powerful lightning storms.

Figure 219 - Friction in pipes can cause sparks.

Discharges such as this, or those between charged and uncharged objects, can cause fires. An example can be between an aircraft and its fuel line. The flow of the fuel acts in the same way as the cloth, charging the fuel pipe. When touching the airplane, which also insulated because it rests of rubber tires, there could be a discharge of the excess electrons. This could be disastrous. As a result both the pipe and the plane are **earthed** so that the excess charge flows into the ground instead of building up.

Most electrical devices are earthed in this way to protect from sparks that could cause fires, or might injure someone.

34.5 The Van De Graaf Generator

It is useful to be able to demonstrate how electrostatic charges behave in a more practical way, and so one of the standard pieces of school laboratory equipment has been developed to provide a charged sphere. This is called a **van der Graaf** generator.

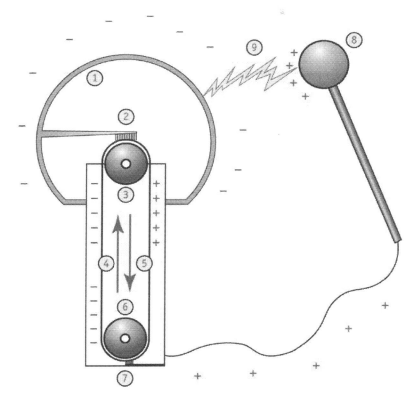

Figure 220 - Van der Graaf generators cause charged particles to build up on the surface of the large globe

1) The hollow sphere (which works best when carefully polished) that stores electrons on its surface
2) Upper brush that collects electrons from the belt (4 and 5)
3) And 6) Teflon (insulator) rollers that the belt is drawn over
4) And 5) the belt, made of synthetic rubber, and carries charge up to 2) where it is deposited on 1)
7) Lower brush that transfers charge onto the belt
8) Discharge ball
9) Spark between the negative sphere and positive discharge sphere.

Numerous fun experiments can be performed, including the classic "hair on end" experiment. This requires a subject to have recently washed, and dry, hair. They are asked to stand on an insulator, such as a polystyrene mat, and place one hand on the sphere. Once the generator is charged, the electrons will seek to move as far away from each other as a result of their repulsive force. They will move into the ends of the hair, and continue to gather there, as far away from the sphere as possible. As a result of the repulsive forces between the hairs, they will lift up and move away from each other.

Figure 221 - Hair sticks on end as the electrons flow to the end of the strands of the hair and repel each other

34.6 Charging by induction.

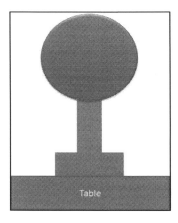

It is possible to induce a charge into a sphere my manipulating the positions of the electrons in an object. It combines the attraction between opposites and a flow of electrons from a different source. In this example a Van Der Graaf generator starts off with no charge on a table, that acts as an insulator.

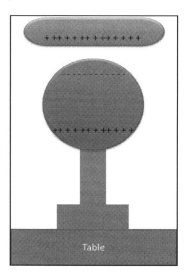

Now bring a positively charged rod near the top of the generator and the electrons in the sphere will move towards the top, being attracted there by the overall positive charge in the rod. These electrons will leave the elements at the bottom of the sphere with a lack of electrons, and so the bottom will end up with a net positive charge.

Keep the rod above the sphere and touch it with a finger, while standing on the ground and you will transfer electrons into the sphere from the ground, replacing those at bottom. When you remove the finger, and rod, the sphere will end up with an excess of electrons. This is an example of **induced charge**.

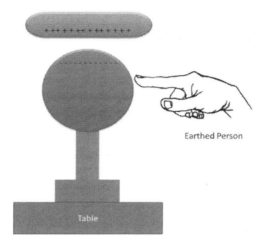

34.7 Unit of charge

Early on in we discussed the need for a common set of measurements when describing quantities. These are called S.I. Units. Strangely, the unit for charge, the Coulomb, named after the famous French Physicist; Charles-Augustin de Coulomb (14 June 1736 – 23 August 1806), is not one of the 7 base units. Charge is considered part of current (see later) instead.

Figure 222 - Charles-Augustin de Coulomb

The unit of charge is called Coulomb (C), but 1C is an enormous charge, and so in most practical cases milli-coulomb (mC) or micro-coulomb (μC) are used instead.

Earlier we discussed that milli- meant 1/1000th and micro meant 1/millionth.

- Charge is measured in Coulomb
- The symbol used for this in equations is Q

34.8 Uses of electrostatic charge

There are times when inducing electrostatic charge can be very useful. In the following examples it can be seen how

Car Painting → Paint droplets are provided with a negative charge, and thus repel each other, spreading out instead of forming larger drops. The car body is given an positive charge, and so the droplets are attracted to the surface of the car, leaving very little waste.

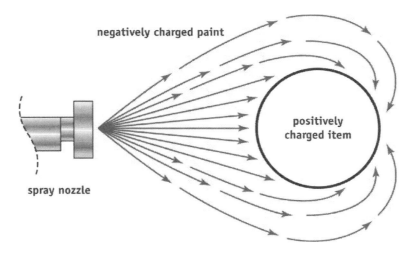

Figure 223 - A positively charged car attracts negatively charged paint particles

Electrostatic Dust Precipitators → As the waste gases, that contain smoke particles, rise up a chimney it passes through a negatively charged metal grid. This deposits electrons on the smoke particles. These, newly charged, particles continue to rise, passing between two positively charged collecting plates. Since opposites attract, the particles will be attracted to the plates. In this way only waste gases that do not contain smoke particles escape into the air.

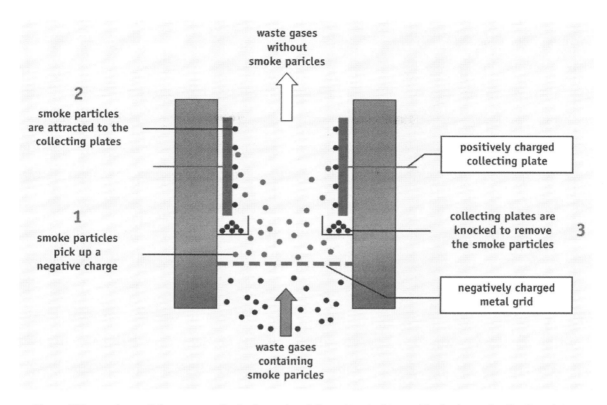

Figure 224 - smoke particles are negatively charged and then attracted to positively charged collecting plates

Inkjet Printers → When the ink passes through the nozzle, it is given a negative charge by the charging electrode. The deflection plates then guide to ink to the right place on the paper by varying how much charge they have. The stronger the deflection plate charge is, the stronger the attraction and therefore the bigger the deflection.

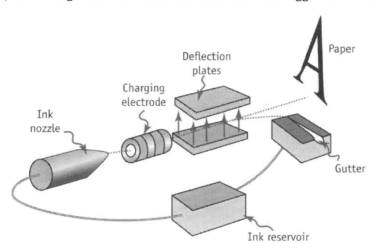

Figure 225 - Negatively charged ink droplets are directed by deflection plates to the right location on paper

Photocopiers → When a document is placed on the glass on the photocopier, a bright light is reflected back off it onto a positively charged copying plate. Where the light falls on the plate, the positive charge is removed, leaving only the dark areas of the plate charged. A negatively charged toner powder is then attracted to the charge areas of the plate, lining it with toner. A paper is then placed on top of the plate and the toner is transferred to this instead. The paper is then heated and the toner is fused to the paper, creating a photocopy of the original.

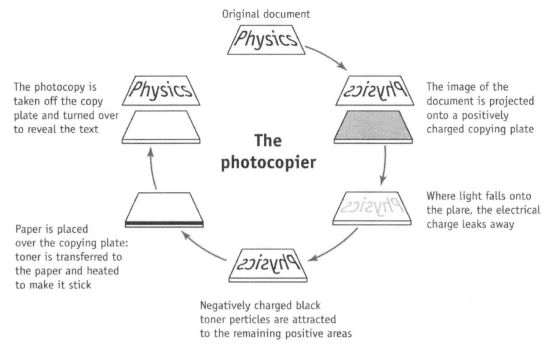

Figure 226 - How static charge attracts toner to the right location to form an image in a photocopier

Questions

1) Give an example of where electrostatic charge might be a hazard.
2) How can the build-up of charge be prevented?
3) On the right, a charged metal can.
    a. Copy the diagram
    b. Draw any induced charges on the can
    c. Why is the can attracted to the rod even though the net charge on the can is zero?
    d. In which direction is the electron flow when your finger touches the can.
    e. What type of charge is left on the can after it has been touched?
4) Explain why a balloon becomes negatively-charged after being rubbed through your hair.
5) Why does this balloon stick to the ceiling?

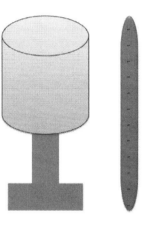

Chapter 35. Electricity

Before getting into the nitty-gritty bit of electricity, it important to understand that electricity involved the movement of energy, using charge as it method of carrying the energy. In other words, charge (or electrons) is capable of carrying energy – a bit like small buckets capable of transferring energy.

35.1 Current

We initially discussed the concept of charge moving as being a current of charge, in the same way that a current in the sea is a current of water.

Current is defined as:

- The Flow of Charge (Q) passing a point each second
    - Measured in Amps, or Amperes – symbol A
    - The symbol used in equations is I (Capital i)
- It is possible to measure the flow of charge using an **Ammeter**
    - Ammeters are designed to plugged in in **series**
    - Ammeters **do not** affect the current

In the following circuit the Ammeters will read the same value since the same number of electrons pass through each one each second. It does not matter that they have passed through the bulb, just that they are flowing around the circuit. In a circuit of this design, called a **series** circuit, the electrons pass from one circuit component to the next until they return to the cell (or battery).

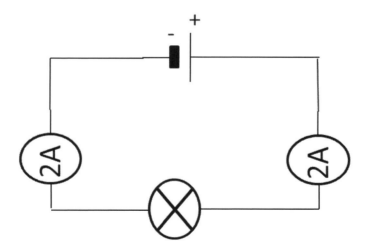

- In a series circuit, the current is the same through the whole circuit.

Charge and current

Since charge and current are closely related, it should not be a surprise that there is an equation that connects the two.

$I = Q/t$	<table><tr><th>Quantity</th><th>Name</th><th>Units</th></tr><tr><td>I</td><td>Current</td><td>Amps (A)</td></tr><tr><td>Q</td><td>Charge</td><td>Coulomb (C)</td></tr><tr><td>t</td><td>Time</td><td>Seconds (s)</td></tr></table>

- 1 coulomb/sec = 1 Ampere
- 2 coulomb/sec = 2 Amperes etc.
- If a current of 2 A flows for 3 seconds, the charge delivered is 6 coulombs.

Current direction

Electrons are repelled by the negative terminal of a cell (or battery) and attracted to the positive terminal. So electrons, as negative particles, flow from **negative to positive**, as in the circuit below. However, when electricity was initially discovered and that it involved the flow of particles, they defined the flow as being from **positive to negative**, as if the particles flowing were **positive** particles, i.e. not electrons. It was only after the discovery that the electron was negatively charged that they realised their error.

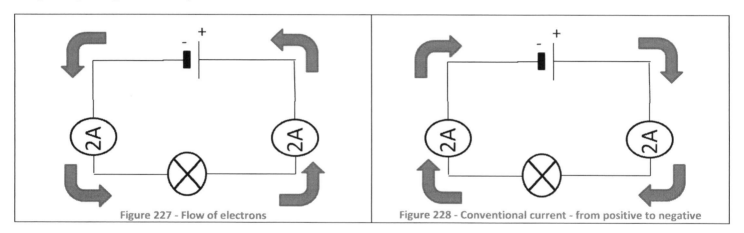

Figure 227 - Flow of electrons

Figure 228 - Conventional current - from positive to negative

The conventional current flows from the positive terminal to the negative terminal of a cell (or battery), the opposite direction to that of the electrons

However, rather than change all the work they had developed, they formed a new concept called the **conventional current**. They defined this is terms of *what would (not what does)* happen *if* a positive particle was flowing in the circuit. In conductors, all positive particles are tied up tightly in the nucleus and do not move, but this was not understood in the early days of the discovery of electricity.

35.2 Questions:

1. Convert these currents into Amperes:

        a. 500mA   (b) 30mA   (c) 4mA   (d) 260mA
    2. What charge is delivered if?
        a. A current of 10A flows for 5 seconds.
        b. A current of 250mA flows for 40 seconds?
    3. Calculate the current in mA if:

Q = 75C   t = 45s
Q = 24C   t = 125s
Q = 130C   t = 25s

    a. Calculate the time in seconds if:

Q = 340C   I = 24A
Q = 70C   I = 3A
Q = 42C   I = 30A
Q = 8C   I = 0.4A

35.3 Voltage

As current shows the number of electrons (or flow of electrons) passing a point each second, the voltage shows the amount of energy per electron (or large group of electrons).

- Voltage is defined to the amount of energy per electron (or per unit charge) in a current
    - It is measured in Volts (V)
    - In an equation the symbol V is used
- When this energy is used to "push" the current around the circuit, it is known as the **Electromotive Force, or EMF**.
- When the amount of energy per unit charge has changed between two points in a circuit, it is known as the **Potential Difference**.
- The Potential Difference is measured using a **Voltmeter**
    - Voltmeters are ALWAYS placed in parallel, across the circuit components that are drawing energy from the circuit – e.g. a bulb, resistor or a bell

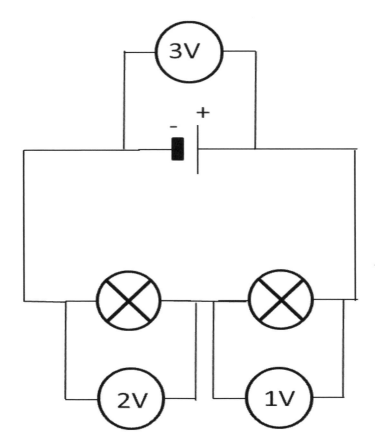

In the circuit drawn above the cell adds 3V of energy per charge to the circuit. When the charge returns to the cell, it will have utilised all that energy. In this circuit, it is the bulbs that are converting that energy to light and heat. This particular circuit has been designed around two different bulbs that are drawing different amounts of energy. It is likely that the left bulb is brighter since it is converting 2V to heat and light, whereas the right hand bulb is only converting 1V.

Notice the following in the circuit: The total of the potential differences across the bulbs is the same as the electromotive force produced by the cell.

The equation that related charge, voltage and energy is:

$V = E/Q$	Quantity	Name	Units
	V	Voltage	Volts (V)
	E	Energy	Joules (J)
	Q	Charge	Coulomb (C)

35.4 Questions:

1) In the circuit below two bulbs are of different sizes and brightness.

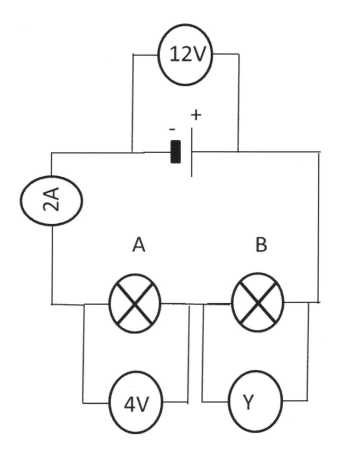

a. What type of meter is meter y?
b. What is the reading on meter y?
c. How much energy does each coulomb have as it leaves the battery?
d. Draw in the conventional current direction.
e. How much energy is lost by each coulomb passing through bulb A?

2) How much energy is transferred by a battery of e.m.f 4.5V when 1C of charge passes through it?

3) Make some changes to the circuit to the right
   a. Add an arrow to show the conventional current direction.
   b. Add a voltmeter to measure a voltage across A.
   c. Add a switch that controls C only.
   d. Add an ammeter to measure the main circuit current.

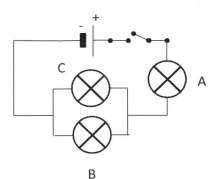

Chapter 36. Types Circuits

We introduced the idea of plugging in ammeters in **series** and voltmeters in **parallel** above. Now we will explore these two concepts in more detail.

36.1 Series circuits

When electrons flow in a circuit and they have to pass through one component after another, it is said that these components are in series. Look at the circuit from the perspective of the electron. If an electron passes through one component, then another, and so on, until it returns to the cell, those components are in series.

Because the electrons pass through one component after the other, its energy is shared between those components. This sharing might be equal, but most often, in real circuits it will vary, depending on the component and how carefully it has been manufactured, or by design.

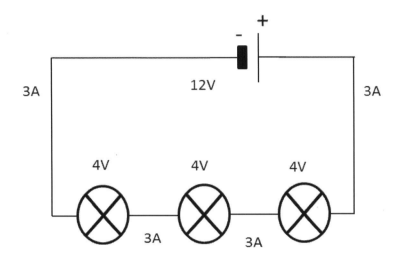

Figure 229 - In the series circuit above, the electrons share their energy between the bulbs, but the current does not change

Two bulbs in series will thus share the energy passing through them and so will be dimmer than one bulb by itself. Also, because the electrons have to flow through one bulb to get to the next and return to the cell, if either bulb breaks, both bulbs will stop working.

- In series, electrons pass through one component, then the next. They have no choice
- If one component in a series circuit fails, the circuit fails
    - It is not possible to control bulbs individually in a series circuit – if one is turned off, all go off
- The components share the energy between each other – not necessarily equally
    - This means that bulbs connected in series become dimmer than one along
- The current in a series circuit is the same everywhere
- There is one benefit of series, and that is that they use less wire than a parallel circuit
- Examples of use is in Christmas trees, when one bulb breaks you have to replace it, and so it is not suitable for a house.

36.2 Parallel circuits

When a circuit is built such that electrons may select different routes from one terminal, back to the other, it implies that there several routes for them to take. These routes, because they start and end at the same place, a considered to be **parallel**. They might not be drawn in a way that looks parallel, but if an electron has a choice NOT to go one route, but go a different route, is will be a parallel circuit.

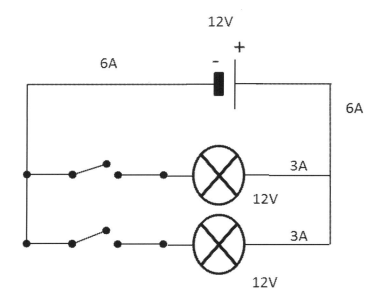

Figure 230 - In the parallel circuit above, the electrons can choose two routes back to the cell.

The implication is that the energy carried by the current is only used in the branch of the circuit that charge flows through. Additionally, since there are several routes from the two terminals, if one component on one of the branches is broken, the other branch can continue to function normally

1. Circuit components each have THEIR OWN CIRCUIT from the battery.
2. Each component has the FULL voltage FROM THE BATTERY.
3. Bulbs in a circuit glow BRIGHTLY.
4. If one bulb is removed, the bulbs in their own circuit STAY ON - e.g. house circuits, lights in cars.

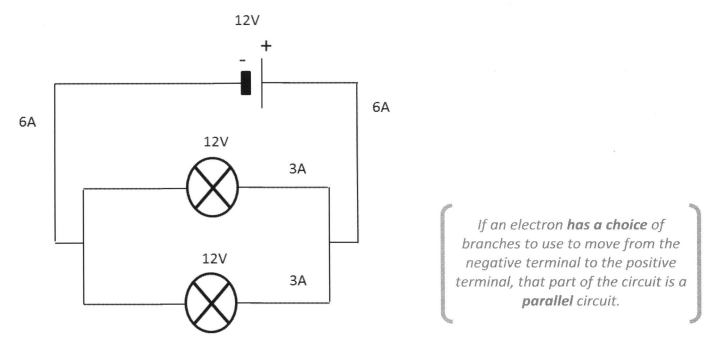

*If an electron **has a choice** of branches to use to move from the negative terminal to the positive terminal, that part of the circuit is a **parallel** circuit.*

The Potential Difference across components is the **same** but the total current in the main circuit is the **sum** of the current in the branches. The current through each branch can, however, differ, depending on the components within each branch. If one branch has two bulbs in series, the current through that branch will be less, but the potential difference across the whole branch will still be the same as across the other branch.

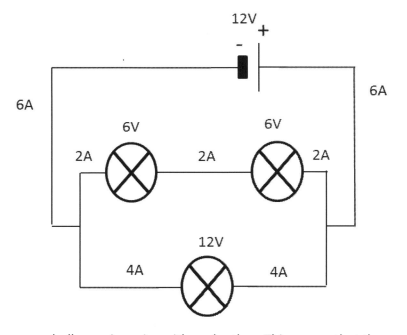

In the circuit above the two upper bulbs are in series with each other. This means that the energy of the charge flowing through them is shared equally between them. The total of the two potential differences adds up to the E.M.F. provided by the cell (6V + 6V = 12V). Those two bulbs are in parallel to the lower, 12V, bulb. Because there is only one bulb in that branch of the circuit, the energy in the charge flowing through that bulb is completely used by it for light and heat.

Note that the currents through the two branches are different, but add up to the total current from the cell (2A + 4A = 6A). We will see later why this is, but for now it is best to consider that the two bulbs in series makes it more difficult for the charge to flow through them than it is to flow through the single bulb.

36.3 Questions:

1) Give two advantages of connecting bulbs to a battery in parallel.
2) The circuit below shows bulbs and ammeters.
   i. What are the reading on ammeters X and Y?
   ii. What is the PD across each of the bulbs?
   iii. Place a switch in the circuit that will turn off both bulbs?
   iv. Place a voltmeter to measure the voltage across bulb a?

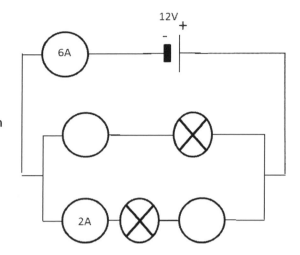

3) The circuit shows a battery connecting a switch and 3 bulbs B1, B2 & B3.
a. Add an arrow to show the conventional current direction.
b. A voltmeter to measure the voltage across B1.
c. A switch that controls B3 only.
d. State and explain what effect adding another cell to the battery would have on the lamps.

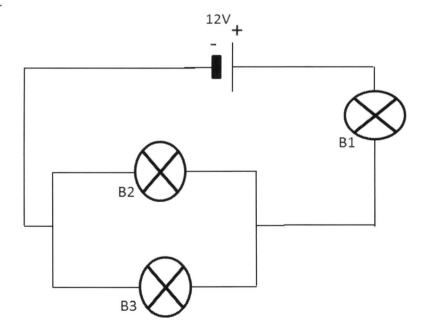

4) Write down the current flowing through:
a. Lamp C.
b. Lamp E.

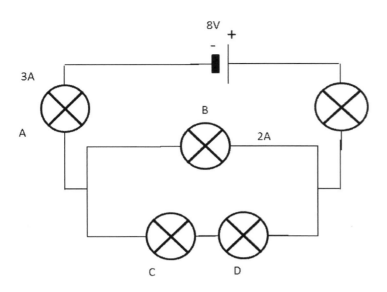

5) How much energy is transferred by a battery of EMF 4.5 V when 1.0 C of charge passes through it?

Chapter 37. Resistance and Ohm's law

37.1 Resistance

Current was described as the rate at which charge flows through a conducting material. It helps considering what is actually happening inside a conductor when charge is flowing. A charged particle is moving between atoms, finding a route from one end of the wire to the other. In travelling along the conductor it will encounter many different routes, most of which involve being repelled by atoms and bouncing around. The "bouncing" will slow down the current, and different materials will cause this to occur more than others.

Scientists look at this at a high level and think of the "bouncing" as resistance to the flow of current.

Resistance is defined as:

- The degree to which current is reduced when flowing through a conductor.
    - Different materials have different resistances
    - Copper has a **low resistance** and so is a **good conductor**
    - Nichrome wire (an alloy of both nickel and chrome) has high resistance
- It is measured in OHMS(Ω) - Ω is he Greek symbol for Omega and is used instead of a western letter

When a potential difference is applied across a resisting material, a current will flow through it. The rate at which the current flows is resisted according to the following equation:

$V/R = I$	Quantity	Name	Units
	V	Voltage	Volts (V)
	R	Resistance	Ohms (Ω)
	I	Current	Amperes (A)

Figure 231 -Georg Ohm - Using equipment of his own creation, Ohm found that there is a direct proportionality between the potential difference (voltage) applied across a conductor and the resultant electric current

This equation shows how, as the resistance (R) increases, the current flowing for a potential difference is reduced. The equation is known as **Ohm's Law**, named after the German physicist, Georg Ohm, and is more commonly written as:

$$V = IR$$

A conductor can have a number of factors that can influence how resistive it is. The longer a wire is, the more atoms there are to bounce off, and so the resistance increases the longer a wire is. A wire that has a larger cross section allows the charges to flow easier, and so a wire that has a larger diameter has a lower resistance. Different materials have different amounts of "free" electrons that may flow, and so have different resistances, e.g. copper has a large "sea" of free electrons that may flow, whereas Nichrome has a lot fewer and so resists the current more.

When a conductor warms up, the atoms inside it vibrate more violently. The result of this is that they take up more volume each, and, as mentioned in thermal expansion, they actually do occupy a larger space. This makes it more difficult for the charge to flow past and so hot conductors have higher resistances

- Resistance increases with the length of a conductor
- Resistance decreases with the cross-sectional area of a conductor
- Different materials are differently resistant (called resistivity)
- Hotter conductors are more resistive

37.2 The heating effect of resistance

As electrons move through any electrical component it will heat up as a result of the charge transferring some of its energy to the atoms in the component. These atoms start to vibrate faster, and this vibration, is heat. This is a direct result of the resistance and the continuous flow of electrons in through the component. It might be useful to consider this as a form of friction between the electrons and the material they are flowing through.

This heating effect of resistance can be used in practical circumstances:

- Toaster heat elements
- Water heaters, such as kettles or immersion heaters
- Electrical grills

- Induction ovens
- The hearing elements in hair-dryers
- "old style" filament light bulbs

This heat is also often the cause of "wasted" energy. For instance, when a filament light bulb is heated up so that it glows, very little of the energy is transformed to light; most is lost as heat. In transferring energy long distances, a lot of the energy can be lost in the wires. It is therefore better to use a high voltage than a high current when going across long distances (see later under transformers).

37.3 Different types of resistors

Type of resistor	What it does	Symbol
Fixed resistance resistor	Provides a known resistance that impedes the flow of current	
Variable resistor	Also known as Rheostats and are used to vary the current flowing through and thus the potential difference across it also (see potential dividers later). Used in HiFis to control the volume.	
Thermistor	Have a **high resistance** when it is **cold** and a **low resistance** when it is **hot**. These are used to control air conditioners and ovens.	
Light Dependent Resistor (or LDR)	These have **high resistance** in the **dark** and a **low resistance** in light. These can be used to automatically control street lights	
Diode	Allows current to flow in **only** on direction once it has reached 0.7V.	
Fuse	A wire in a resistor will melt when too much current is passing through it	

37.4 Questions

1) In a resistor there is a PD of 240V with 3A flowing through it. Calculate resistance.
2) In a resistor with a resistance of 70Ω, 12A. Calculate PD across it
3) In a resistor with a PD of 12000V across it and a resistance of 36000 Ω. Calculate current passing through it
4) Name 2 other components which use the heating effect of resistance.
5) Name 2 reasons why tungsten wire is used in light bulb filaments.

In a normal conductor, CURRENT will vary with PD, if the TEMP of the conductor is KEPT THE SAME.

37.5 Ohm's law

At school you might be asked to explore Ohm's law using the circuit outlined below. In this circuit the variable resistor will change the amount of current passing through the series circuit AND will also use some of the PD provided by the cell. The amount of PD that is left for the fixed resistor depends on how much the variable resistor used. This is covered in more detail further down in "potential dividers".

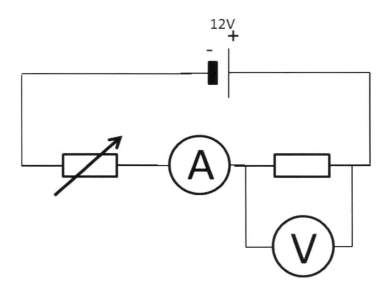

Figure 232 - Circuit to explore Ohm's law

The practical will involve choosing a number of equal settings for the variable resistor and creating a table similar to the one below that records different currents based on the PD across the resistor.

Current/A	Voltage (Potential Difference)/V
0A (no current)	0
0.5	5
1	10
1.5	15
2	20
2.5	25

The table itself shows how, if the voltage doubles, the current also doubles, and it thus proportional, but this not the end of this practical. You will need to produce a graph. This will be a straight line, where the gradient of the line is the resistance, per the equation $V = I \times R$.

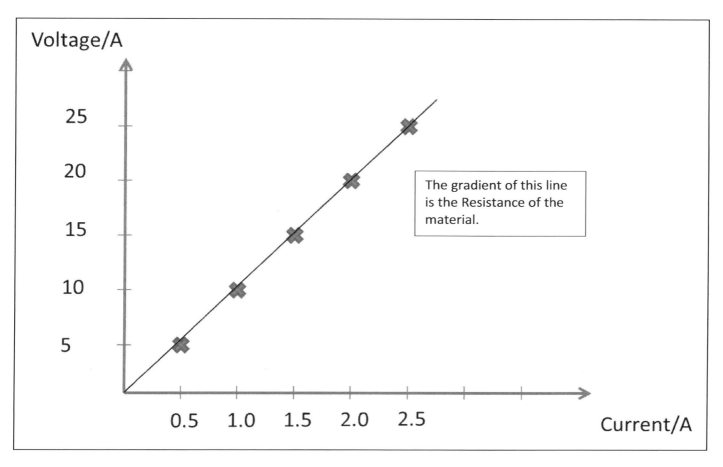

Metal conductors obey ohm's law provided their temperature is constant. These are called Ohmic conductors and their graphs will be as above.

37.6 Current and PD graphs

It is an important part of the syllabus to be able to understand how these graphs are interpreted and how they vary between three different resistors.

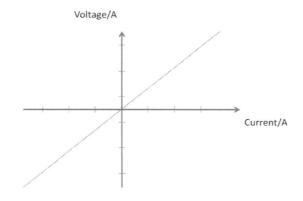

Figure 233 - Simple "Ohmic" response of a resistor to a potential difference across it

Firstly, the **fixed resistor** shows Ohmic behaviour with a directly proportional relationship between the PD across the resistor and the current flowing through it. The graph below shows this with both positive PD across the resistor, and the reverse.

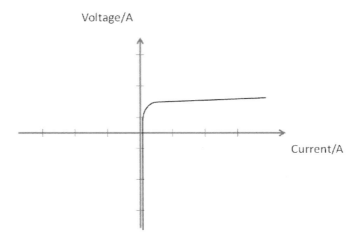

Figure 234 - How the voltage changes the current in a diode

Secondly, we look at the **diode**. This specialist type of resistor is used to ensure that current can only flow in one direction, once a certain PD has been reached, and could be used to protect components from current flowing the wrong direction. We would expect a negative PD to allow no current through and a very low resistance once this threshold has been reached.

Thirdly, we look at a **tungsten filament light bulb**. In order to understand this graph, it is vital to understand what happens in the filament. When current flows through tungsten it heats up dramatically. It becomes so hot that it glows and it is this glow that is used to provide light. Another side effect of this is that, as the tungsten gets hotter, its resistance increases. This is because the hot atoms are vibrating more violently and therefore occupy, on average, more space, making it more difficult for the electrons to flow. As such, we should, initially, see that it behaves as an Ohmic material, and then that the resistance would increase, causing the gradient of the line to rise.

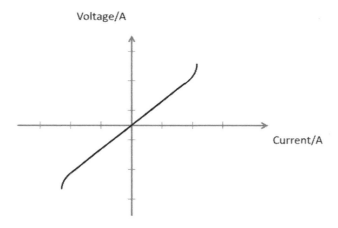

Figure 235 - The change in gradient of the voltage vs current graph shows how, when a bulb is hot, the resistance increases.

Some exams will swap the current and voltage axes, and then the graphs will look slightly different:

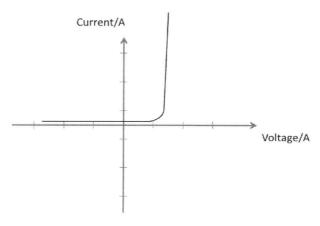

Figure 236 - this is the more common way of looking at how a diode works. When the voltage reaches +0.7V, the diode lets the current through

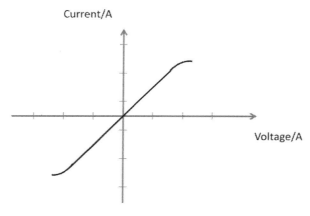

Figure 237 - This diagram shows how, when the light bulb is hot, the resistance increases dramatically, making it difficult for the current to flow.

37.7 Questions:

1) The graph lines on the graph are for 2 different conductors.
   a. Which conductor has the higher resistance?
   b. Add a third line to show a very low resistance.
2) A resistor has a resistance of 8Ω.
   a. If the current is 2A, what is the PD?
   b. What PD is needed to produce a current of 4A?
   c. If the PD falls to 6V, what is the current?
3) Calculate the resistance of the Tungsten wire at:
   a. 1500°c (v = 2v, I = 1A).
   b. 3000°c (v = 2v, I = 3A).
4) In the diode graph, does the diode have its highest resistance in the forward or reverse direction? Explain your answer.
5) In a current-carrying wire, name 3 ways to reduce the effect of resistance.

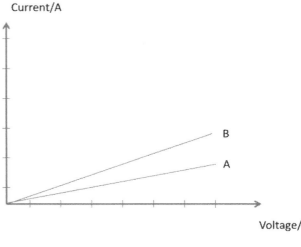

Chapter 38. Resistors in circuits

When circuit components that resist current, e.g. fixed value resistors, they interact either by increasing the resistance (in a series circuit), and thus reducing the current and reducing the amount of energy available for the rest of the circuit, or by reducing the resistance (in a parallel circuit), thus making it easier for current to flow, and allowing more energy in each branch.

38.1 When placed in Series

- When two, or more, resistors are placed in series, as in the circuit diagram below:
    - They increase the potential difference across the two resistors
    - They increase the resistance by adding to each other
    - They reduce the current flowing through that part of the circuit

The total resistance of a circuit is calculated using the following equation:

$R = R_1 + R_2 + ... R_n$	R is the total resistance. R_n are the value of each resistor in series.

The following shows some examples of resistors:

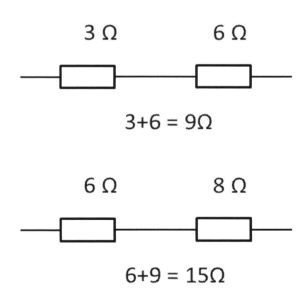

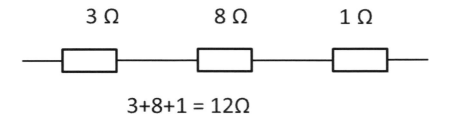

3+8+1 = 12Ω

38.2 When placed in parallel

- When two or more resistors are placed in parallel:
    o The total resistance is reduced to lower than the lowest resistor
    o The current passing through the parallel circuit is increased
    o The potential difference is the same across the branches

There a special case where it is easy to calculate the total resistance:

Same Resistance
- When 2 resistors placed in parallel of the same resistance their combined resistance is HALF A SINGLE RESISTANCE.
- When three resistors are placed in parallel, that are the same, the total resistance is ONE THIRD OF A SINGLE RESISTANCE

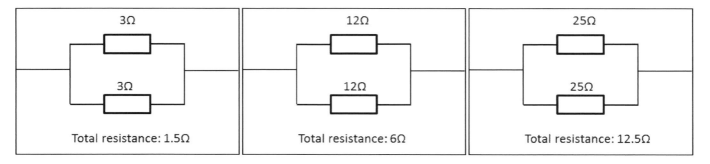

| Total resistance: 1.5Ω | Total resistance: 6Ω | Total resistance: 12.5Ω |

Different resistances
Based on laws that are covered in A-level physics, and called Kirchoff's laws, it can be shown that there is a simple relationship between the total resistance across a parallel circuit and values of the resistors.

When 2 or more resistors placed in parallel with DIFFERENT RESISTANCES, their combined resistances can be found using the equation:

| $\frac{1}{R_T} = \frac{1}{R_1} + \frac{1}{R_2} + \cdots + \frac{1}{R_n}$ | R is the total resistance. R_n are the value of each resistor in series. |

Examples

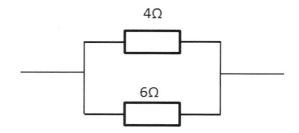

$$\frac{1}{R_T} = \frac{1}{4} + \frac{1}{6} = 0.416. \; R_T = \frac{1}{0.416} = 2.4\Omega$$

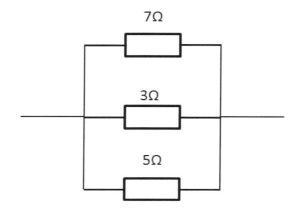

$$\frac{1}{R_T} = \frac{1}{7} + \frac{1}{3} + \frac{1}{5} = 0.67 \; \therefore \; R_T = \frac{1}{0.67} = 1.5\Omega$$

When combining series and parallel sets of resistors:
With different parallel resistances in a different circuit to the series, analyse the resistors to find out which are operating in parallel and which are operating in series.

In the example below, we consider the two resistors in parallel to be connected in series with the third resistor. To work out the total resistance we first calculate the total resistance in the parallel circuit and then add that resistance to the last resistor.

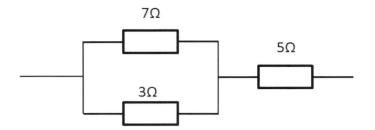

$$Total\ Resistance = R_P + R_S$$

Where R_p is the parallel resistance and R_s is the series resistance (5 Ω).

$$\frac{1}{R_P} = \frac{1}{7} + \frac{1}{3} = 0.48 \quad \therefore R_P = \frac{1}{0.48} = 2.1 \Omega$$

$$R_T = 2.1 + 5 = 7.1 \Omega$$

In the following example, the combination shows a set of resistors in series, within a parallel configuration. In this case we will first calculate the series resistance, and then work out the parallel resistance.

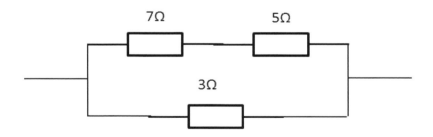

$$Series = R_S = 7 + 5 = 12 \Omega$$

$$Parallel = \frac{1}{R_T} = \frac{1}{12} + \frac{1}{5} = 0.283$$

$$\therefore R_T = \frac{1}{0.283} = 3.5 \Omega$$

38.3 Questions:

1. Analyse the circuit below:
   a. What does the ammeter read?
   b. What is the PD across each of the resistors?

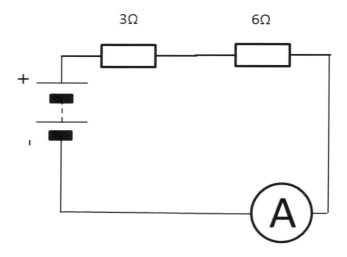

2. Calculate the resistances below:

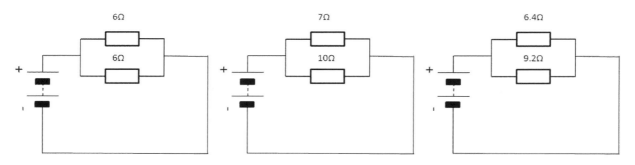

3. Calculate the resistances and current below:

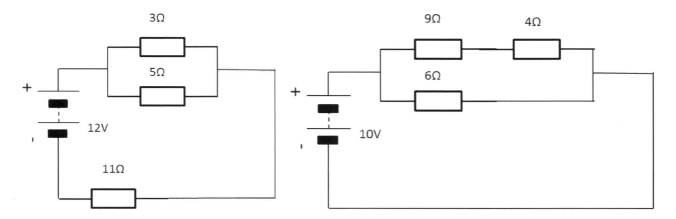

4. Which resistor arrangement. A or B has the lower resistance? Check your answer by calculation.

A

10Ω

B

990Ω
10Ω

Chapter 39. Electrical energy

39.1 Energy transferred

Electricity is all about the transfer of energy through the movement of charge in a conductor. It is important that we remember what the definition of current (I) and potential difference or voltage (V) is:

1) Current can be thought of as the number of charged particles passing a point in a wire each second
2) Voltage is the amount of energy associated with each charged particle.

We already discussed the difference between "conventional current" and "electron flow", and noted that one flowed in the opposite direction to the other. However, for the concept of transferring electrical energy, the direction is less important at iGCSE and GCSE than it is for A-level.

Imagine, if you will, that an electron passes each second with 1J of energy associated with it. That would mean that 1J/s, or 1W, will be passing each second. If we have 2000 electrons passing each second, each with 1J, it would be 2000W. 1J is a huge amount of energy for one electron to be associated with, so in most electrical calculations it is better to use the concept of charge, measured in Coulomb. One Coulomb each second is equivalent to 6×10^{18} electrons passing by each second.

We can then redefine the current above as:

1) Current is the number of coulombs passing each second, or I=Q/t

Following this idea again for voltage, instead of the amount of energy in one electron, we consider it as the amount of energy in one Coulomb, or the amount of energy per 6×10^{18} electrons. So we can redefine voltage from above as:

2) The amount of energy per Coulomb of charge, or V=E/Q

So, if we have a certain number of Coulombs passing a certain point, each containing a certain amount of energy, we end up with a certain amount of energy passing that point each second.

$$P = \frac{Q}{t} * \frac{E}{Q} = \frac{E}{t} \text{ or } P = V * I$$	

Quantity	Name	Units
P	Power	Watts (W)
Q	Charge	Coulomb (C)
t	Time	Seconds (s)
E	Energy	Joules (J)
V	Voltage or Potential Difference	Volts (V)
I	Current	Amps (A)

To work out the amount of energy that has flowed over a period of time, we just need the amount of energy each second, and multiply it by the number of seconds.

$$E = P * t = V * I * t \text{ and since } Q = I * t, E = V * Q$$

Quantity	Name	Units
P	Power	Watts (W)
t	Time	Seconds (s)
E	Energy	Joules (J)
V	Voltage or Potential Difference	Volts (V)
I	Current	Amps (A)
Q	Charge	coulomb (C)

Worked Example

If a 12V heater is switched on for 60s and uses a 2A current, how much thermal energy is produced?

Use the equation:

$$E = P * t = V * I * t$$

List what is known:

I = 2A
V = 12V
t = 60s

Substitute in the values into the equation

E = V*I*t = 12 x 60 x 2 = 1440J.

Worked Example

A coil has a resistance of 5Ω and the current through it is 0.6A. Calculate the heat energy produced by the coil in 20s.

Use the two equations

$$E = P * t = V * I * t$$

And

$$V = I * R$$

Combine the equations:

$$E = I^2 * R * t$$

List what is known:

I = 0.6A
R= 5Ω
t=20s

Substitute into the equation

$$E = 0.6^2 * 5 * 20 = 36J$$

39.2 Electrical power

Power is the concept of how rapidly energy is transformed. A lot of energy transformed quickly would require a lot of power; similarly, little energy transformed over a long time would imply a small source of power. This is briefly discussed earlier.

In electricity power is defined as

Definition: The RATE at which ELECTRICAL ENERGY is TRANSFORMED.

$P = \dfrac{E}{t}$	Quantity	Name	Units
	P	Power	Watts (W)
	E	Energy	Joules (J)
	t	Time	Seconds (s)

Figure 238 - James Watt, the Scottish inventor after whom the unit for power is named. One of the instigators of the industrial revolution

- Power is measured in WATTS (W), or Joules per second.

Worked example

A toaster supplies 3000J/s, so its power is 3000W. Most electrical devices require large quantities of power to perform their tasks, and so electrical appliances have a POWER RATING measured in watts of KILOWATTS (kW).

1kW = 1000Watts.

For example:

Kettle = 2400W (2.4kW)
Iron = 1100W = (1.1kW)
Lamp = 60W

39.3 Electrical power in circuits

In circuits, because VOLTAGE and CURRENT are involved, the power equation is:

$$P = V * I$$

Quantity	Name	Units
P	Power	Watts (W)
V	Voltage or Potential Difference	Volts (V)
I	Current	Amps (A)

Consider what this means:

- Each electron "contains" a certain amount of energy (the Voltage)
- A certain number of electrons passes each second (the current)

39.4 Questions

1) Current = 0.1A, PD = 400V. Calculate Power
2) PD = 24V, Power = 240W. Calculate Current.
3) Power = 1.5kW, Current = 3A. Calculate Potential Difference

39.5 Power in resistors

As electrons pass through and between atoms, some of their energy is transferred in those interactions to those atoms. This causes the atoms to vibrate more vigorously, thus taking more space, and making it more difficult for the electrons to pass. The electrons lose some of their energy and so their potential energy is less after the interaction. The result is that the media it is passing through gets warmer, increases in resistance and that a potential difference exists between the two ends.

- Due to the HEATING EFFECT of RESISTANCE through a resistor, ELECTRONS LOSE POTENTIAL ENERGY. This is called THERMAL EHERGY.
- Energy is **dissipated** in the resistor.

Formula for calculating power dissipation in a resistor:

	Quantity	Name	Units
$P = I^2 * R$	P	Power	Watts (W)
	R	Resistance	Ohms (Ω)
	I	Current	Amps (A)

Worked examples

What power is dissipated in a 5Ω resistor when the current through it is (a) 2A (b) 4A?

    a)   P = I² R = 2² x 5 = 20W.
    b)   P = I² R = 4² x 5 = 80W.

Note: DOUBLING the current produces FOUR TIMES the POWER DISSIPATION – this becomes very important when we discuss the transfer of electricity over long distances (see "transformers" below)

39.6 Questions

1) A hairdryer takes 10,000J of energy from the mains supply in 5s. Calculate its power in (a) watts (b) kilowatts.
2) An electric heater uses 4A connected to a 230V supply. Calculate its power.
3) A light bulb is rated at 36W. It is connected to a 12V supply. Calculate the current through it.

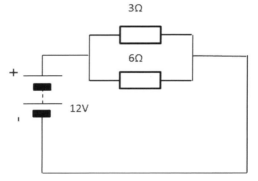

4) Examine this circuit diagram and:
a) Calculate the current in each resistor.
b) Calculate the power dissipated in each battery.
c) Calculate the power of the battery.
d) If a battery of twice the PD was used, what would its power be?

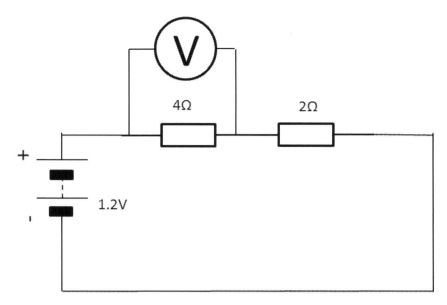

5) Examine this circuit diagram and:

Calculate:

   a. The combined resistance of the 2 resistors.
   b. The current flowing through the cell.
   c. The current in the 4Ω resistor.
   d. The reading of the voltmeter.
   e. The power produced in the 4Ω resistor.

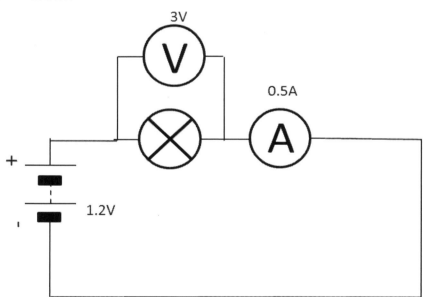

6) Examine this circuit diagram and:
   a. Calculate the resistance of the lamp.
   b. What is the power of the lamp?
   c. Calculate the power if the potential difference was doubled.

Chapter 40. Electrical safety

In order to protect the user from electrical shocks, which could be lethal, or sparks, which could cause fires, there are a number of simple devices that are used to keep users and their property safe:

40.1 Circuit Breakers

A circuit breaker is a simple electromagnetic device (see later) that triggers when too much current is passing through. If too much current passes through, the magnet in the circuit breaker briefly becomes excessively powerful and opens a switch, stopping the current from flowing. If the switch is opened, it is very easy to reset it by moving a lever.

Circuit breakers are typically placed in electrical cupboards and are tuned to work for specific loads – e.g. for the kitchen, one for the upstairs and one for the downstairs in a house.

40.2 Fuses

Fuses are very simple devices that have a wire inside them. One part of the fuse is connected to the live part of a power supply, and the other end connects to the appliance. If too much current passes through the fuse, the wire heats up and melts, thus stopping current from flowing.

The symbol for a fuse is:

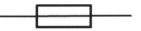

The line going through the centre of the resistor represents the fuse wire. If a fuse burns out, it must be replaced. This usually involves unscrewing a plug, prying out the old fuse and replacing it.

Figure 239 - Typical fuses

40.3 Fused plugs

The United Kingdom and many other countries have fused plugs. These have three copper pins that are plugged into an electrical wall socket. Each part of the plug is designed to provide certain extra levels of security.

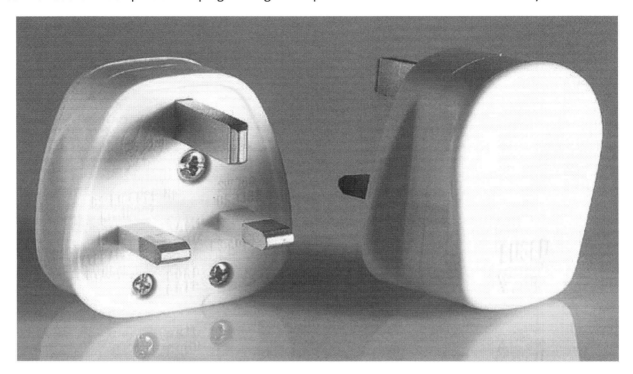

Figure 240 - A UK three-pini mains socket plug

The power cable enters the plug from beneath, and when opened, we can follow the cable and note the safety features that are designed into the plug.

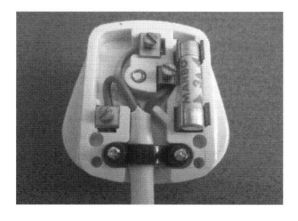

Figure 242 - Safely wired

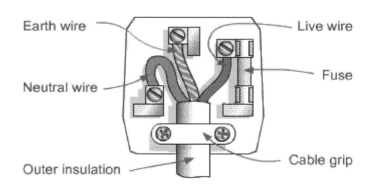

Figure 241 - Locations of the wires and parts of a plug

40.4 Safety features

1) Live wire → Colour-coded. Brings LIVE VOLTAGE to plug – usually brown, but can be red
2) Earth wire → SAFETY WIRE. Connect plug to earth. Excess current to EARTH.
3) Neutral wire → Completes the circuit. Kept at ZERO VOLTAGE.
4) Plastic casing → ELECTRICAL INSULATOR.

5) Fuse → Placed in the LIVE WIRE. Wire inside melts with excess current.
6) Cable grips → Holds cable firmly in place.
7) Connecting screws → Holds wires in place.
8) Double insulation → Has 2 plastic casings around main cable to plug.

40.5 Hazards

HAZARD	DANGER
Damaged Insulation	Possibility of electric shock.
Cables overheating	Excess current in thin cables causes plastic to catch fire.
Damp conditions	Water is an excellent water conductor.
Short-circuit	Electric current travels along path of least resistance.
Switches	Live current present if accidently switched on.

40.6 Calculations used in electrical safety

One of the topics that is covered in most syllabi is that of selecting the right fuse. In order to make this calculation you need to know what the maximum running current that a device will have and then select the right fuse.

The steps in doing this are:

1) Calculate the typical running current within the device, e.g. using P=VI or V=IR.
2) Select the next highest rated fuse, e.g. if you calculate the device needs 4A, use a 5A fuse.

Worked Example
A kettle is rated at 2kW, what rating of fuse will you need? 1A, 3A, 5A or 13A fuse.

First we need to know the voltage, and all houses in the UK have a supply of 230V. We then use the equation:

$$P = V * I, reorganise\ for\ I = P/V$$

List what we know:

V= 230V
P= 2kW = 2000W

Substitute into the formula:

I=2000/230=8.7A

The nearest, higher, fuse is 13A. So we choose that one. If we had chosen 5A, the fuse would constantly burn out.

40.7 Earthing

Electrical circuits may be connected to ground (earth) for several reasons. In mains powered equipment, exposed metal parts are connected to ground to prevent user contact with dangerous voltage if electrical insulation fails. Connections to ground limit the build-up of static electricity when handling flammable products or electrostatic-sensitive devices.

This can, for instance, be very useful when static electricity might build up through friction between objects. One particularly dangerous situation might occur when planes are being refuelled at an airport. As the liquid flows into the plane, there is friction. This causes a charge to build up. Adding an earth connection to the airplane and the tanker ensures that no sparks between the two objects occur.

40.8 Double insulation

If it is possible to ensure that no live components will ever be exposed to a user of an electrical appliance, this is preferable to earthing. In many devices, such as drills, hair dryers, washing machines and so on, the concept ensures that:

1) The wires are connected internally, within the machine, so no live parts of the wire are exposed
2) The device itself is made of an insulator (or is encased in an insulator, but it might be hidden underneath a conducting shell).

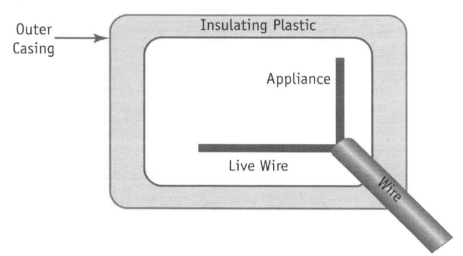

Figure 243 – A double insulated appliance has the appliance made of an insulated material and the cables are never exposed

The symbol used on an electrical device to show that it is has double insulation are two squares, one encased in the other:

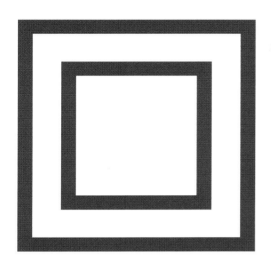

Notice on the right-hand image how the symbol is shown on the label of an electrical device.

40.9 Dealing with electrical accidents

If something goes wrong at home and someone has a shock, here is what you need to do:

1) **DO NOT TOUCH THE VICTIM** - you may also receive an electric shock.
2) Turn off the main switch immediately. This will either be located in the meter box or inside your home, away from the meter box. The main switch turns off the mains electricity supply to your home. If you have a second supply of electricity this will need to be disabled too.
    a. If it is not immediately possible to shut off the power, move the victim away from the source with a non-conductive item, like a stick or blanket.
3) Dial 999 (If you have a mobile phone dial 112 to connect to Emergency Services. Contact your provider for details about your service).
4) Do not touch the victim until you are sure that they are no longer in contact with the electrical current.
5) If the area is safe, check the victim for a response, breathing and pulse. If necessary start resuscitation (**you will need training to do this, you can get this from St. John's Ambulance, or a scouting group**).
6) Keep the victim warm.
7) **Do not touch burns**, break blisters or remove burned clothing. Always seek medical advice after an electric shock, even if injury is not apparent.

Figure 244 - Before touching a victim, turn off the power

Chapter 41. Effects of electromagnetism

41.1 The generator

Since electrons are miniscule magnets that can be manipulated by magnetic fields, it is possible to make the electrons move along a wire using a magnet. We have already learned that electrons moving along a wire is called current. Thus it is possible to cause a current by moving a magnet **across** a wire, or, a wire **across** a magnetic field.

When a current is caused in this way, it is called an "induced current".

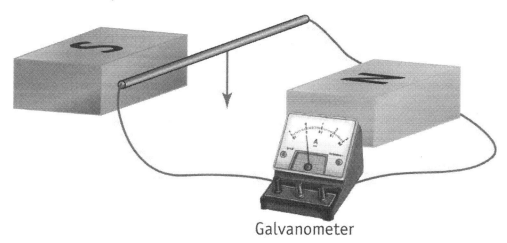
Galvanometer

Figure 245 - Moving a wire between two magnets will induce a current in the wire

In the diagram above, a wire is passed down, between to opposite poles of two magnets. Invisible field lines connect the two poles, as discussed above. These lines are called flux lines. As the wire moves through them, it "cuts" the field lines. As a result, a current will flow through the wire. This can be detected with the galvanometer – a very sensitive ammeter.

- Magnetism is used to produce electricity
- The movement of a wire across (cutting the flux lines) a magnetic field **induces** a current

This process is an energy transformation process where:

Kinetic Energy (the movement of the wire) -> Electrical energy

The effect is known as **electromagnetic induction**. The amount of induced current can be increased by increasing the number of flux lines cut per second, or the rate of cutting flux lines.

Since the density of the flux lines is increased when the strength of the magnet is increased, a stronger magnet will increase the current. Moving the wire across them faster will increase the rate at which the lines are being cut, and so also increases the current. Having a longer magnet with a longer wire going through it will also increase the number of flux lines cut each second.

Current can be increased by:

1) Moving the wire faster.
2) Using a stronger magnet.
3) Increasing the length of wire.

41.2 How to find the direction of the induced current

From the diagram above, with the wire passing through a magnetic field, it is clear that the whole physical process of passing a wire through a magnetic field and generating a current is a 3 dimensional process. The wire and the magnetic field are at **right angles**, the wire and the movement are at **right angles**, and, finally, the movement and the magnetic field are at **right angles**. Since the current is flowing through the wire, the current is at right angles to both the magnetic field, and to the movement.

Sir John Ambrose Fleming, of Lancaster, England, devised a simple way of working out how the directions of these three physical actions relate to each other.

Figure 246 - John Ambrose Fleming 1849-1945

Fleming's Right-hand Rule
This is used in generators, and is thus known as the generator rule. Since most people are right-handed, you can try to remember this as the hand that makes things happen, or generates things.

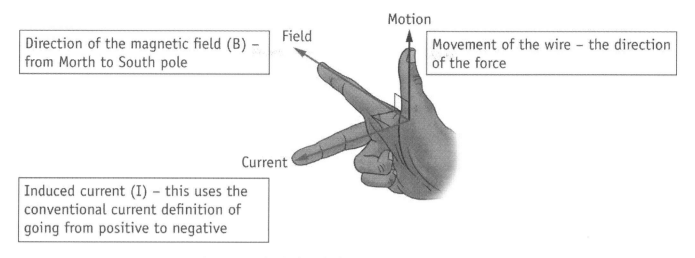

Direction of the magnetic field (B) – from North to South pole

Movement of the wire – the direction of the force

Induced current (I) – this uses the conventional current definition of going from positive to negative

Figure 247 - Fleming's Right-hand rule, or the generator rule

In physics, the magnetic field is denoted by the symbol **B**, current by the symbol **I** and a force (in this case the direction of the movement of the wire) with an **F**.

You can remember this rule using the acronym "FBI", as in the American FBI (Federal Bureau of Investigation).

- Thumb, first finger and second finger must be at **RIGHT-ANGLES** to each other.

41.3 Force on a conductor in a magnetic field

For a conductor in a magnetic field with a current flowing through it, the force experienced is proportional to the product of the length of the wire in the field, the current and the magnetic flux density (B).

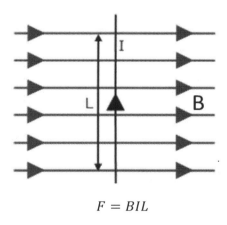

$$F = BIL$$

Induced **current** (I) – this uses the conventional current definition of going from positive to negative

electromagnets, using coils allow us to magnify the effects of magnetism due to a flow of current; a coil increases the effect of a current induced by a moving magnetic field.

A coil is formed by winding a wire around and around some kind of tube, e.g. a toilet roll, or a piece of PVC tubing.

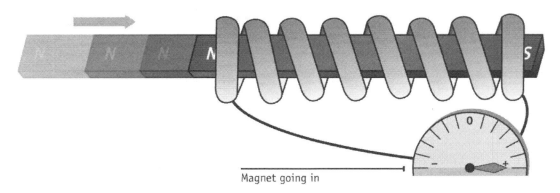

Figure 248 - Step 1, the magnet it pushed into the coil

In step 1, the magnet is pushed into the coil. The movement of the magnet, and its associated magnetic field, past the coils causes the magnetic field lines (flux lines) to be cut by the wires. The cutting of the flux lines by the wires induces a current in one direction.

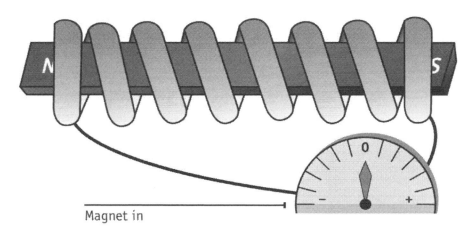

Figure 249 - Step 2, the magnet is left stationary inside the coil. No current is induced

In step 2, the magnet is brought to rest. The flux lines are stationary; they are not actively being cut by the wires. Since they are not being cut by the wires, there is no current being induced, as shown by the galvanometer.

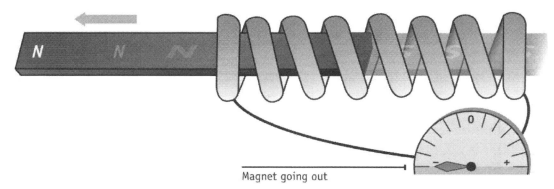

Figure 250 - Step 3, the magnet is pulled out. Current is induced in the opposite direction to that in step 1

In the third step, the magnet is pulled out. The flux lines are being cut again, but in the opposite direction. This causes a current to flow, but in the opposite direction to that in step 2.

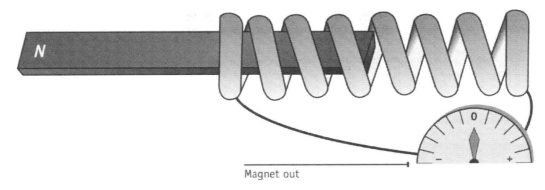

Figure 251 - Last step, the magnet is outside the coil, at rest, and no current is thus induced.

41.5 Observations

In the previous section of this book we noted that the wire moving inside a magnetic field will induce a current in the wire. However, here the magnet is moving instead:

- A moving, or varying, magnetic field within a coil of wire will induce a current in the coil (we will see later, under transformers, that it is not even required for either the magnet, or the coil, to move, just for the magnetic field to be varying)
- The direction of the current changes depending on the direction the magnet is moving through the coil
- The magnet (or coil) must be moving for a current to be induced.

We mentioned that this whole process is an energy transformation process of kinetic energy -> electrical energy. So, if we increase the kinetic energy by moving the magnet (or coil) faster, we will gain more current:

- Increasing the speed of the movement of the magnet will increase the current (and potential)

41.6 Faraday's Law

Figure 252 - Michael Faraday (1791-1867)

Michael Faraday noticed that the rate at which the flux lines were "cut" by a wire influenced how strong the current was. Rephrasing his law:

"The rate at which flux lines are cut is proportional to the induced current"

So, if we increase the density of either the flux lines, by using a stronger magnet, or by having more wire cutting them each second, the current will increase:

- A stronger magnet will increase the current (and potential)
- A longer wire within the magnetic field will increase the current (and potential)
    - This could be by using a coil or by using a long wire (and magnet)
        - Adding more winds to a coil provides more wire within a magnetic field

41.7 Lenz's Law

When a current is induced in a coil, that current in produces a magnetic field. The direction of this induced magnetic field will oppose that of the magnet that induced the current.

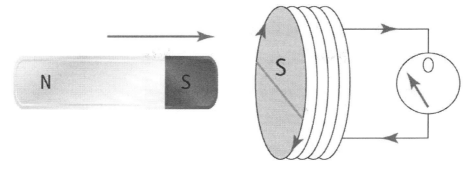

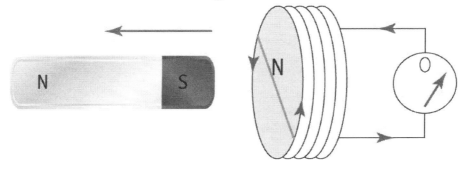

Figure 253 - When the magnet enters the coil, the current will produce a south pole to repel the magnet, whereas, when the magnet is pulled out, the induced current will produce a magnet that attracts, or opposes, the magnet being removed.

Finding the direction of the induced current in the coil
This involves combining the "right hand coil clasp" rule with Lenz's law:

1) The magnet entering the coil will produce pole that opposes its entry into the coil, i.e. it will be the same polarity as the end of the magnet entering it
2) The right-hand clasp coil rule will then show the direction of the current that will produce this magnet.

An induced electromotive force (EMF) always gives rise to a current whose magnetic field opposes the original change in magnetic flux.

Current Flows	Current Flows Opposite Way
Produces a NORTH POLE at the end of the coil.	Produces a NORTH POLE at the end of the coil. Current is flowing in opposite direction.
Current is OPPOSING movement of magnet (By trying to REPEL it).	South pole OPPOSING movement of magnet (By trying to ATTRACT it).

Typical uses of induced currents
1) Old style tape recorders pass a magnetic tape over a coil of wire. This induces a current in the wire that can be amplified and listened to
2) Videos, such as VHS or camcorders, use a similar mechanism, but the induced current is interpreted as a combination of sound and vision.
3) Microphones where the membrane causes a magnet to vibrate inside a coil, inducing a current that is used to transfer the sound to a device for recording, or amplifying

41.8 Questions

Use Fleming's Right hand rule to identify the missing piece of information:

1) Which direction is the current?

2) Which direction was the movement?

Current

3) Which side is north and which is south?

Movement

4) Which is north and which is south?

Movement (out of paper)

Current

5) Use Fleming's Right-Hand rule to calculate the direction of the induced current:

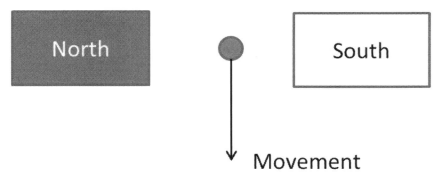

Movement

6) Examine the following diagram:

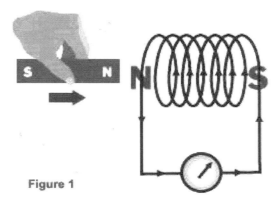

Figure 1

The galvanometer's needle deflects to the right when the magnet approaches.

   a) What happens to the galvanometer reading when the magnet moves away from the coil?
   b) What would have happened with the needle if the magnet had approached with the south pole first?
   c) What would happen to the reading if the magnet had been moved more rapidly
7) List three ways that it is possible to increase the size of the induced current
8) State Lenz's law
   a. Why is a North Pole produced in a coil if a magnet is moved in North Pole first?
   b. Why is a South Pole produced when the same magnet is withdrawn from the coil?
9) Calculate the direction of the magnetic field below and then mark the poles of the magnets

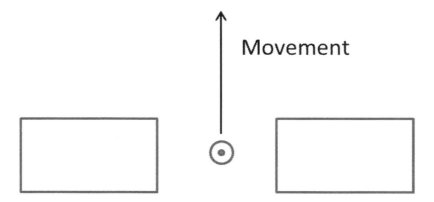

41.9 The Motor effect

When a wire already carries a current, there is a magnetic field around it. The direction of this field is identified using the "Right hand thumb" rule, as shown in **Error! Reference source not found.** on page **Error! Bookmark not defined.**. If this wire, with its current and associated magnetic field, is then placed inside a permanent magnetic field, such as between two opposite poles of a magnet, their respective magnetic fields will interact. The wire will experience a force.

The size of the force is increased by:

1) Increasing the current.
2) Using a stronger magnet.
3) Increasing the length of wire in the magnetic field (this could involve using a coil)

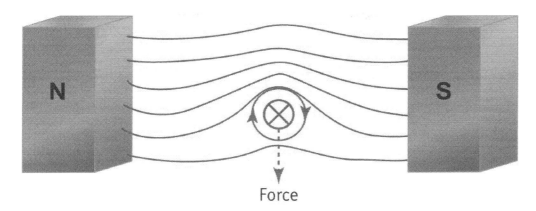

Figure 254 - Force experienced by a wire with its current going into the paper. The flux lines build up on the side where the directions of the two magnetic fields, pushing the wire downwards.

In common with the generator, the three components of this force are all at right angles to each other. John Fleming, again, has provided a tool to work out how these relate to one and other, called "Fleming's Left Hand Rule".

41.10 Fleming's Left Hand rule

In Figure 254, above, it is possible to see how the three components work at right angles to each other. The current is going **into** the paper; the magnetic field is going across the paper, and the force the wire experiences is down the paper. If the wire is loose, it will move in the direction of that force.

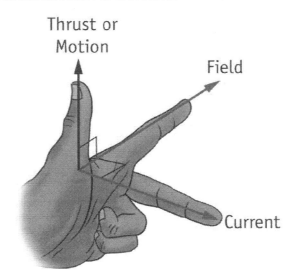

Figure 255 - Fleming's left-hand rule, or the motor effect rule

Please notice that this rule is the "left-hand rule", but is very similar to that of the generator, right-hand rule, shown above.

Note that **NO FORCE** is produced if current and field are in the **SAME DIRECTION**.

Turning effect in a coil
When a current is passed through a loop of wire inside a magnetic field, a turning effect is produced. The force of the current going into the magnetic field is opposite to that of the current coming back out again. This will cause that loop of wire to turn. If the coil is attached to bar that allows it to rotate around it, the electromagnetic induction will cause it to spin, as long as the force is caused to change direction once it has become vertical.

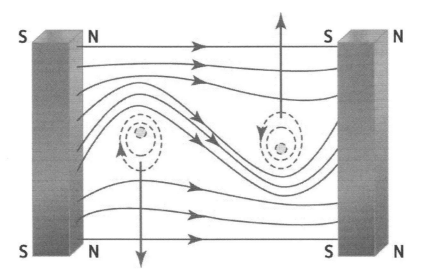

Figure 256 - Turning force on a coil in a magnetic field

A coil, of course, is just a number of continuous loops, and so:

- When a coil is placed between the poles of a magnet, a TURNING EFFECT IS PRODUCED.
- The current flows in OPPOSITE DIRECTIONS along the sides of the coil.

41.11 Questions

Using Fleming's Left-Hand rule, attempt the following:

1. Calculate current

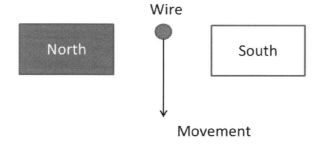

2. Determine the movement that caused this current

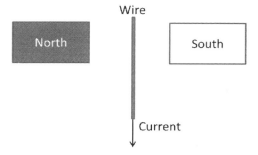

3. Mark on this diagram which pole is which

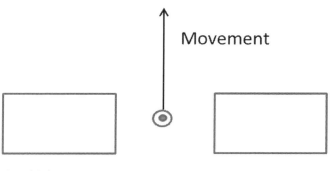

Mark on this diagram which pole is which

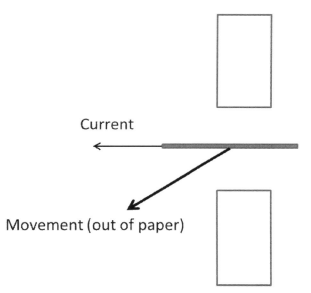

41.12 Making use of electromagnetic induction

Since an electrical current passing through a magnetic field can cause movement, and since a wire cutting across flux lines causes current to flow through the induction of a potential difference across the opposite ends of that wire, electromagnetic induction is a way of providing energy transformation.

For a wire with current flowing through a magnetic field, it is possible to transform electrical energy to kinetic energy, and for a wire cutting through magnetic flux lines it is transforming kinetic energy to electrical energy.

1) A generator uses the kinetic energy of a wire cutting flux lines to produce electricity.

2) An electrical motor uses a spinning coil inside a set of permanent magnets to provide kinetic energy

The AC generators (alternator)

A generator produced a current that changes direction on a regular basis. We have discussed the idea of frequency and how this is used to describe something that oscillates regularly. We learned that this is called the "frequency" and is measured using Hertz, Hz, which means "number of times per second".

A generator produces electricity that changes direction regularly and the rate at which it changes direction is called its frequency. In most households the electricity that is provided is in the form of this **alternating current** (AC) and it has a typical frequency of between 50 and 60Hz.

If we were to plot the current against time we would get a sine curve, similar to the one shown below:

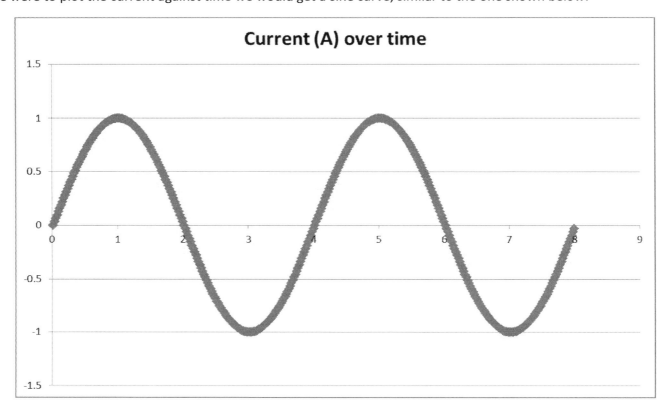

Figure 257 - The graph shows how the current changes direction regularly over time. This is called an alternating current

A simple generator consists of a few simple items:

1) A coil that is attached to a shaft in the centre of it. The shaft allows the coil to spin on its axis.
2) A permanent magnet. The one drawn is a horseshoe type, but they can be separate magnets
3) Slip rings that are part of the same assembly as the coil, and so spin with the coil. As they are made of a conductor, such as brass, they conduct the electricity
4) Conducting brushes that are made of either carbon or strands of copper. These transfer the current from the slip rings to the external circuit

When examining this assembly from the perspective of one of the wires in the coil, it become clear that:

1) When the wire is moving upwards through the flux lines, a current is induced in one direction

a. When this occurs at right angles to the flux lines, the current generated is at its maximum since it is cutting most flux lines each second. This can be seen at times 1, 3, 5 and 7 in Figure 257, above.
2) When it is moving horizontally, across either the top, or bottom of its circle, it is moving along the flux lines and so is not cutting any flux lines.
   a. When this occurs, no current is induced. This can be seen at times 0, 2, 4 and 8 in Figure 257, above, where the current is 0.
3) When the wire passes over its top or bottom, the wire ends up cutting the flux lines in the opposite direction to stage 1, above, and so the current changes direction. This can be seen at times 1, 3, 5 and 7 in Figure 257, above.

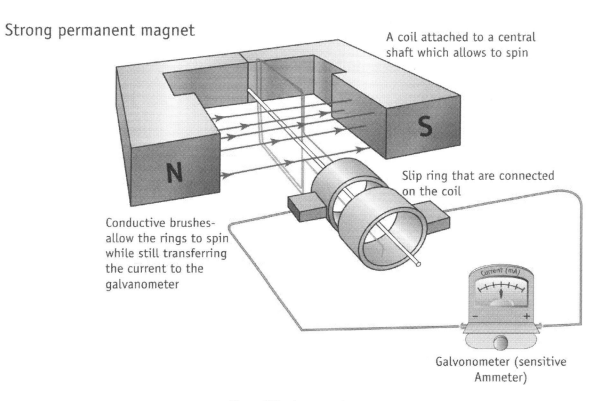

Figure 258 - An generator

The amount of EMF. (and current) can be increased by:

- Increasing the number of turns on the coil
- Increasing the area of the coil.
- Using a stronger magnet.
- Rotating the coil faster.

The DC motor

The transformation of electrical energy to kinetic energy in an electrical motor is performed using DC, or **Direct Current**, current that only flows one way. In **Error! Reference source not found.** we saw how a turning force can be generated by placing a coil inside a permanent magnetic field.

The wires on one side of the coil have a current that has an opposite direction to the wire on the other side. As a result the two sides experience forces in opposite directions.

1) If a coil is carrying a current, the forces on it produces a TURNING EFFECT.

That coil, if attached to a central shaft, will spin until it reaches a vertical position. At this point, the forces are still opposite each other, but because they both pull **outwards** the coil will end up stuck in a vertical position. What is needed is a mechanism that, once the vertical position has been reached, the current changes direction in both wires, so that the forces change direction **inwards**. This is achieved using **split ring commutators**. These commutators are part of the same assembly as the coil, and so spin with them around the shaft. When the commutator is aligned with the coil in a vertical position, the empty gap between the rings is aligned to the brushes and no current flows through the wire, but the spinning motion of the assembly continues through its momentum until the other commutator touches the opposite wire. This change means that the current now changes direction in that wire, and the force reverses, and causes the wire to be pulled in.

2) The split ring commutator reconnects to the opposite wire when at assembly is vertical, causing the force to change direction

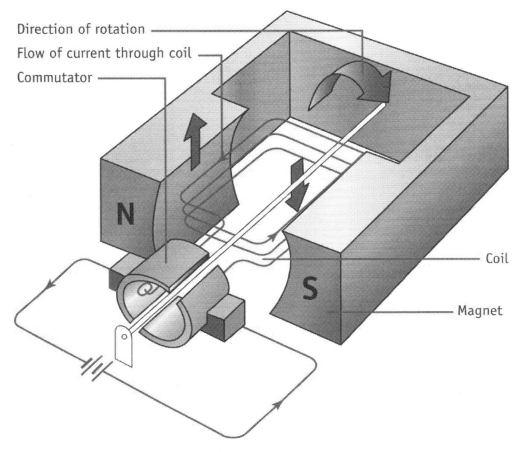

Figure 259 - The motor effect. The commutator rotates with the coil as part of the same assembly. As it rotates, the brushes alternately touch the other commutator, causing the current to change direction in the coil, thus changing the direction of the force in it

In the next figure, follow the motion of the part right of the split ring commutator that is labelled "A".

1) In stage one the wire connected to that side of the split ring commutator is connected to the negative terminal of the battery. The current is flowing through it, out of the magnet and the force is thus upwards. The opposite side of the coil is being forced downwards.

2) In stage 2, the central image, the commutator no longer touches the brushes, but the spinning motion continues due to the momentum of the coil assembly. No force is experienced by the coil since there is no current.
3) In stage 3, the commutator has made contact with the opposite terminal, the current start flowing, but in the opposite direction commutator is now connected to the positive terminal. As a result the force pulls it downwards.

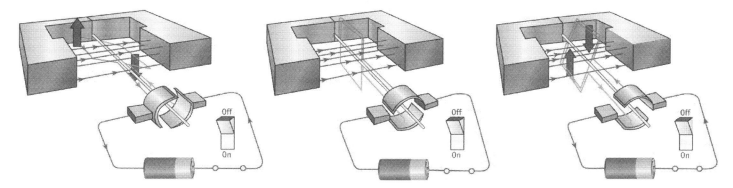

Figure 260 - How the commutator changes the direction of the flow of current

The commutator and coil are both made of a good conductor, such as copper.

Observations
- Forces have maximum turning effect when coil is horizontal.
- The force can be increased by:
    o Increasing the current.
    o Increasing the area of the coil.
    o Using stronger magnets.
    o Increasing the number of turns of wire on the coil.

41.13 Questions

1) Below is a diagram showing the end view of a coil in a simple electric motor.

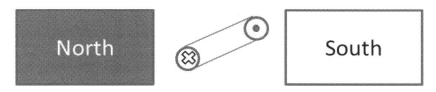

   a) Redraw the diagram to show the coil's position when the turning effect on it is
      (i)   Maximum
      (ii) zero.
   b) Use the Fleming's Left-Hand rule to work out which way the coil will turn.
2) List 3 ways to increase the current from an A.C. generator.
3) What is the coil's position when the current is at its maximum in a generator?
      a. Why is the current maximum in this position?
      b. A wire is placed between the poles of a magnet. When a current flows in the wire, a force acts on it causing it to move.
4) Use Fleming's Left-Hand rule to find the direction of the force on the wire. Show this on the diagram with an arrow labelled F.

a. State what happens to the force on the wire if
b. The size of the current is increased.
c. The direction of the current is reversed.
d. A weaker magnet is used.

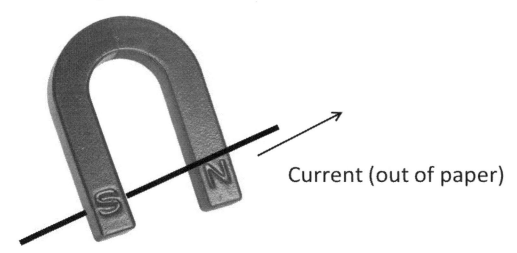

Chapter 42. Transformers

When an alternating current runs through a wire it induces a magnetic field around that wire. Unlike the DC version of this, the magnetic field changes with the direction of the current. This means that there is a changing magnetic field around the wire. Although not immediately obvious why, another wire sat next to this one will experience this varying magnetic field and will have a current induced at exactly the same frequency as the first wire.

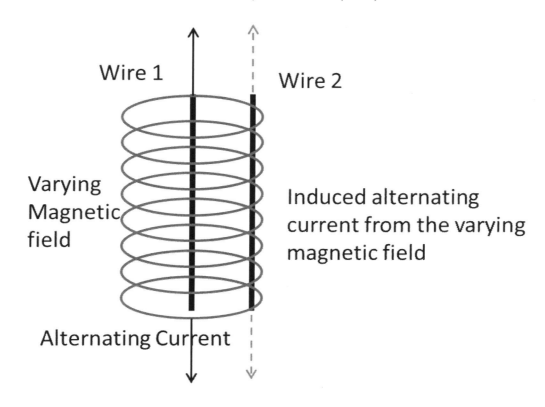

Figure 261 - a varying, or changing, magnetic field will induce a current in a neighbouring wire

This process of current in one inductor inducing a voltage in another nearby inductor is called **mutual induction**.

The use of this is not immediately obvious. However, it is used all through the world in devices call "Transformers". A transformer takes energy from a power source, such as a power station and changes it so that it is more efficient to transmit that energy across long distances.

In order to do so, it reduces the current through the wire, but increases the amount of energy per unit charge within that current, by increasing its voltage. In Chapter Chapter 39, Electrical energy, we introduced the concept of power dissipated in a wire as:

$$P = I^2 R$$

The longer the wire, the bigger the resistance, and so a very long wire, such as that from a power station to a city, would have a large resistance. Engineers design these wires to reduce that resistance as far as possible. But the largest cause of power dissipation is that contributed by the current since the power is proportional to the **square** of the current.

All energy supply is about the transfer of power, and the underlying equation for this is:

$$P = IV$$

So, if we reduce the current, but increase the voltage, it is possible to deliver the power more efficiently. The way of doing so is to **step up** the voltage across long distances. However, high voltages are VERY dangerous, and so once the power is provided near a city, it is important to **step down** to a safer level.

Transformers are used extensively to perform this task.

A transformer uses the physics phenomena shown in Figure 261, above, but with three enhancements.

1) Coils are used instead of a single wire stretched out
2) An Iron core is added to connect the two wire coils – an iron core is a soft magnetic material, so it will easily magnetise and then demagnetise
3) Different numbers of winds of the coil are used at each end depending on whether it is a **step up transformer**, or **step down transformer**.

Transformers are also designed very carefully so that there is as little power loss as possible between the coils. There is, however, no way to completely make the stepping process 100% effective.

	Quantity	Name	Units
$V_P * I_P > V_S * I_S$	V_p	Voltage across primary coil	Volts (V)
	I_p	Current through primary coil	Amps (A)
	V_s	Voltage across secondary coil	Volts (V)
	I_s	Current through secondary coil	Amps (A)

However, for the purposes of your GCSE, or iGCSE, **you can assume that they are the same**.

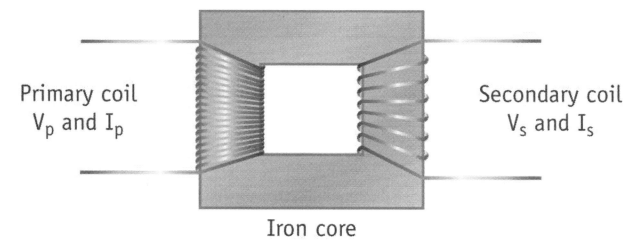

Figure 262 - A step down transformer - notice how the primary coil has more winds than the secondary coil

In the next figure, the primary coil is switched to the side with the lower number of winds. This makes it a **step-up** transformer.

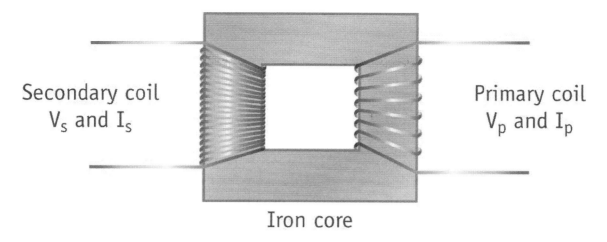

Figure 263 - Step down transformer

Transformer Equation

The relationship between voltage and number of turns can be expressed using the transformer equation:

$$\frac{V_p}{V_s} = \frac{N_p}{N_s}$$

Quantity	Name	Units
V_p	Voltage across primary coil	Volts (V)
N_p	Winds in primary coil	Number of winds
V_s	Voltage across secondary coil	Volts (V)
N_s	Winds in secondary coil	Number of winds

Worked example:
A transformer transforms 240V AC to 12V AC for a model railway. Calculate the number of turns on the secondary coil of the primary coil has 1000 turns.

So:

$$\frac{V_p}{V_s} = \frac{N_p}{N_s}$$

$$\frac{240}{12} = \frac{1000}{N_s}$$

Therefore $N_s = \frac{12 \times 1000}{240} = \frac{12000}{240} = 50\ Turns$

42.2 Questions

1) A transformer has a turn ratio of 1:4. Its input (primary) coil is connected to a 12V AC supply.
    a. Calculate the output voltage.
    b. What turns ratio would be required for an output voltage of 36 volts?
2) The diagram shows a simple transformer.

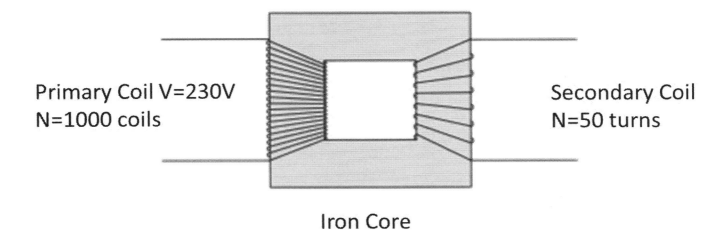

Primary Coil V=230V
N=1000 coils

Secondary Coil
N=50 turns

Iron Core

    a. Is this a step-up or step-down transformer?
    b. Calculate the output voltage.
3) Explain the term "Mutual Induction."
4) A transformer has 22000 turns on its input coil with a voltage of 6000V. The voltage on the secondary coil is 18000V.
    a. Calculate the secondary coil turns.
    b. Give a turns ratio,
    c. Is this step-up or step-down?
5) A power station produces electricity at 25000V. This is increased to 400,000V by a transformer.
    a. Use $V_2/V_1 = N_2/N_1$ to calculate the number of turns on the secondary coils if $N_1 = 2000$.
    b. Calculate a turns ratio.
    c. Is this a step-up or step-down transformer?

Power through a transformer

Transformers are used to transfer power at different voltages from one part of the energy grid to another in order to:

1) Reduce the loss of energy to heat in long cables
2) Keep users safe from high voltages

Transformers are designed to perform this task very efficiently, but, as mentioned above, there is *some* energy loss albeit it less than 5%. This means that the input power is very close to that of the output wire.

At GCSE level you might be asked questions about how much is lost, but most calculations are based on working out a missing value in the **power conservation** equation:

	Quantity	Name	Units
$V_P * I_P = V_S * I_S$	V_p	Voltage across primary coil	Volts (V)
	I_p	Current through primary coil	Amps (A)
	V_s	Voltage across secondary coil	Volts (V)
	I_s	Current through secondary coil	Amps (A)

If no energy is wasted and the transformer is 100%, the power will be the same on both output and input coils.

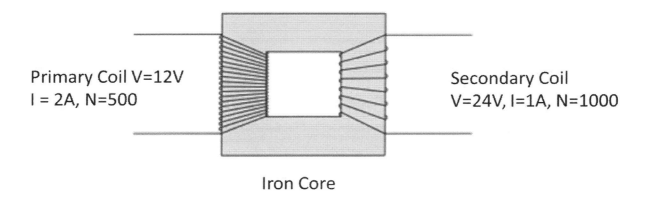

Iron Core

Worked example

In the transformer above, the primary coil has 12V AC across the terminals, and a current of 2A passing through it. The primary coil has 500 winds and the secondary has 1000 winds. Calculate the current through the secondary coil.

Firstly we use the transformer equation to work out the voltage across the secondary coil:

$$\frac{V_p}{V_s} = \frac{N_p}{N_s}$$

V_p= 12V
N_p=500
N_s=1000

Rearrange and solve for V_s

$$V_s = \frac{V_p N_s}{V_p} = 12 * \frac{1000}{500} = 24V$$

Now use the power conservation equation from above

$$V_P * I_P = V_S * I_S$$

$V_p = 12V$
$I_p = 2A$
$V_s = 24V$

Rearrange and solve for I_s

$$V_s = \frac{V_p I_p}{I_s} = \frac{12 * 2}{24} = 1A$$

The current through the secondary coil is 1A

Transformer efficiency

Transformers are not 100% efficient as a result of their physical design. Engineers constantly seek ways to improve on these.

- RESISTANCE (coils are not perfect electrical conductors) – this causes the transformer to heat up
- EDDY CURRENTS → currents are induced in the iron core by the changing magnetic field
- FIELD LINE LEAKAGE → energy is wasted as field lines in the primary coil may not cut those in the secondary coil.

Worked Example

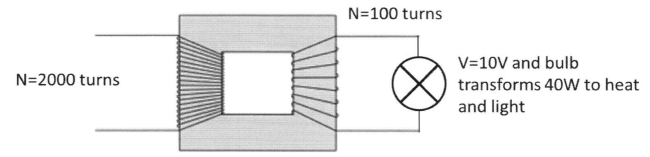

Assume that the transformer configuration above is 100% efficient; calculate the supply voltage and current in the primary coil

First use the transformer equation and rearrange to calculate the primary voltage:

$$\frac{V_p}{V_s} = \frac{N_p}{N_s}$$

$V_s = 10V$
$N_s = 100$
$N_p = 2000$

$$V_p = \frac{V_s N_p}{V_s} = 10 * \frac{2000}{100} = 200V$$

Primary voltage (V_p) is thus 200V.

Now use the power conservation equation from above

$$V_P * I_P = P_S$$

V_p= 200V
P_s=40W

Rearrange and solve for I_p

$$I_p = \frac{P_s}{V_p} = \frac{40}{200} = 0.2 A$$

42.3 Power across the country – the national grid system

All countries produce electrical power at diverse location, often far away from cities for security purposes, or because the power stations are eye-sores. This makes the transmission of that power very important to the communities that rely on it.

1) The power supply should be reliable
2) The transmission should be as energy efficient as possible
    a. Not only is the power dissipation lower in the long cables because of the low current/high voltage combination, but thinner wires can also be used, saving on money and resources.
3) More recently it has been decided that energy production should also be provided using a mixture of different sources, e.g. wind turbines, nuclear power stations and fossil fuel power stations.

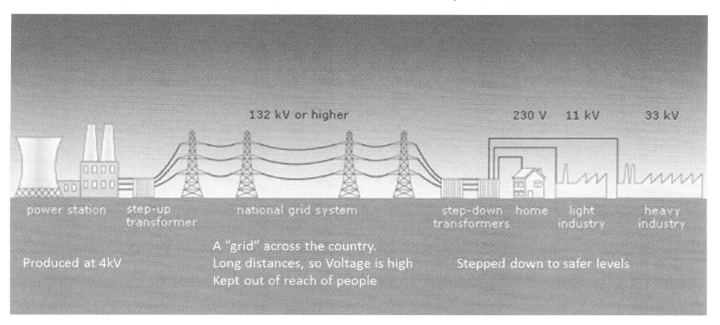

Figure 264 - Using transformers in energy transmission to reduce loss through heating in the wires

Local distribution of power

Figure 265 - Large substation

All power (except for solar cells) is produced using generators and as such are based on an Alternating Current (AC). As outlined above, the power is delivered at high voltages to decrease the loss through heat in the long-distance cables. Once the cables reach a conurbation, a **substation** reduces the voltage to more manageable levels using **step-down transformers**.

These substations then distribute the power to more local substation that are located within suburbs and are contained within small buildings, often close to a main road.

A large substation might step the voltage down to a few kV, and then the local substation steps that down again, to 230V, for use in homes.

Note that only AC current can be stepped up and down in this way. DC current will NOT work with transformers. DC can be transmitted across long distances using completely different methods called arc valves, but this is first covered at university level.

Long distance transmission of power

There are issues involved in transmitting energy across long distances. Power stations are often built in rural areas and the power lines must pass through areas that are considered to be beautiful. High voltage cables are extremely dangerous and so must be kept away from causing danger to people and animals. Two options are available

Figure 266 - Local Substation

Pylons

These high structures are built to keep the cables high above the ground, out of harm's way. They carry the current in copper wires spun around an aluminium core and are designed to be tough, withstanding the forces of nature. They are, however, considered to be ugly and it is very controversial if they are placed in areas of outstanding beauty.

Figure 267 - Controversial location for a pylon run

- Pylons can be very ugly

Pylons, however, do have the benefit of being fairly easy to repair since they are easily accessible. Any fault is found simply by observing the cable.

- They use lower cost, thinner, cabling
- They use higher voltages, and so lose less energy as heat in transmission
- They are easier to install

Underground cables

These cables can be used in areas where pylons are unsuitable:

- Areas of natural beauty
- Across stretches of ocean floor

They are however harder to maintain and repair and to keep out of harm's way. As a result the voltages are dropped, the current is increased and so thicker cables are used to reduce the resistance, driving up the costs.

- Underground cables are expensive to run
- Use thicker, more expensive, cabling
- Are more difficult to maintain and repair

42.4 Questions

1) Complete the following table

Primary P.D.	Secondary P.D.	Primary Turns	Secondary Turns	Step-up/down
100V AC		10	100	
100V AC		100	10	
240V AC	12V AC	200		
11000V AC	132000V AC		12000	

2) A transformer uses 230V AC mains to power a heater:

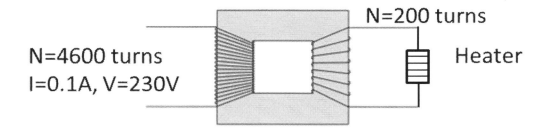

Calculate:

   a. The voltage across the heater
   b. The power supplied by the mains.
   c. The power delivered to the heater.
   d. The current through the heater.

3) Explain:

   a. A transformer will not work on DC.
   b. If a transformer increases voltage it reduces current.
   c. Transformers are never 100% efficient.

d. Electricity is transmitted at high voltage.
4) Calculate the power loss from a cable carrying a 10A current and a resistance of 5Ω.

Chapter 43. Electronics

43.1 Electronic control systems

Most electronic circuits operate using very low voltages and very low currents since they are often based on the use of batteries, or other sources of DC currents, such as DC power adapters. Some of these devices will increase the current internally for further use, such as Op Amps.

Devices that convert electrical signals into another form are called **transducers.**

Control Systems

Control systems are usually represented using simple processing symbols to illustrate, at a higher level, the flow of the information. In more detailed versions of this, such as in computing or electronics GCSE these symbols are explored in much more detail.

In the following example, a microphone is amplified:

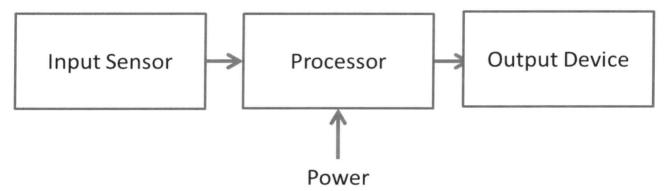

This process flow diagram indicates that there is an input sensor, in this case a microphone, that there is a processing devices, in this case an amplifier, and then, finally, that there is an output device, in this case a loud speaker.

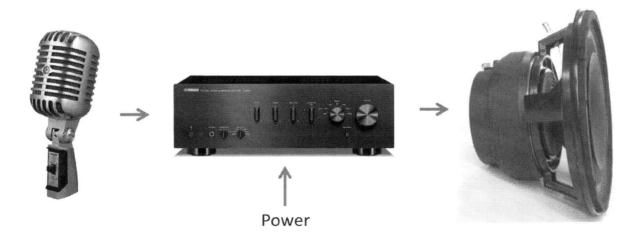

All control systems have an **input** sensor sending signals to a **processor**. This controls the flow of power to an **output** device.

The following table shows a number of devices that can act as input sensors, or output devices.

Table 13 - Example input and output devices

INPUT SENSORS	OUTPUT DEVICES
LDR	LED
Thermistor	Bulb
Capacitors	Bell
Transistors	Motor
Variable Resistors	Relay

Combining these devices provides opportunities to develop all kinds of electronic devices, using a combination of these devices and other circuit components.

43.2 Main circuit components

Many of the following components have been discussed, briefly, before. In this section we will be exploring their purpose within circuits.

In brief, the following are the purposes of some of the many components available for use in a circuit.

- **Resistors** → change current and voltage for other circuit components.
- **Capacitors** → store electric charge.
- **Diodes** → allow current one way only.
- **Transistors** → amplify signals.
- **Relays** → act as electromagnetic switches.
- **Potential Dividers** → Can share the potential difference between parts of a circuit.

Diodes

Diodes only allow the current to pass through in one direction, and are thus used to protect parts of a circuit that might be damaged by current flowing in the wrong direction. They are also used to convert alternating current (AC) to direct current.

- Diodes are drawn as arrows pointing as an arrow in the direction of the **conventional** current

Figure 268 - Circuit symbol for a diode

Figure 269 - Diodes

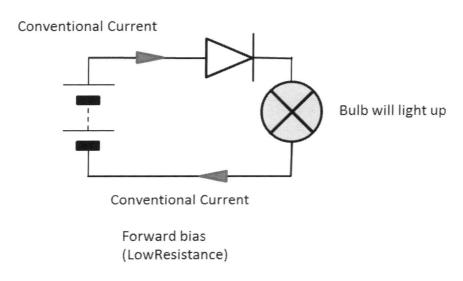

Figure 270 - In this circuit, the current is allowed through the diode and the bulb will light up

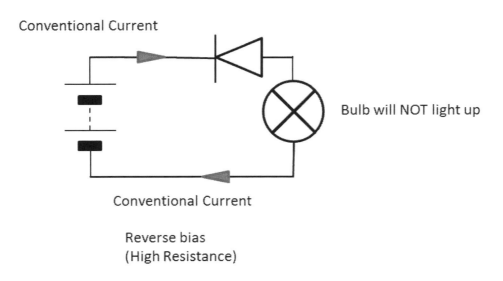

Figure 271 - The diode resists the current as it is plugged opposing the conventional current

Rectifiers

When diodes are used to change AC to DC they are called **rectifiers**. Many devices require DC instead of AC, e.g. **motors, TVs, Computers, Mobile Phones, Battery charges**, etc.

The process of changing AC to DC is called **Rectification**.

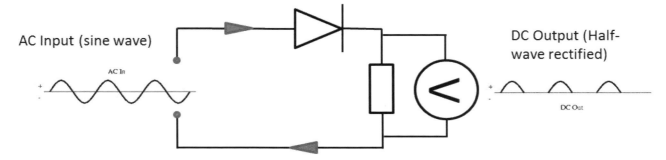

Figure 272 - in this circuit, only the current that flows in the direction of the diode is let through. The positive peaks are the ones left

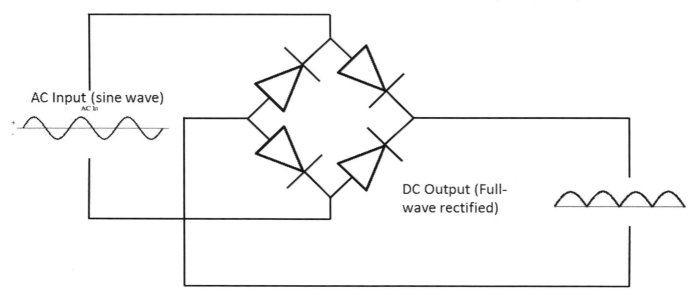

Figure 273 - In this configuration, the output signal is output as a fully rectified DC link, but it still is not smooth

A capacitor can be added between the two outputs to smooth the signal:

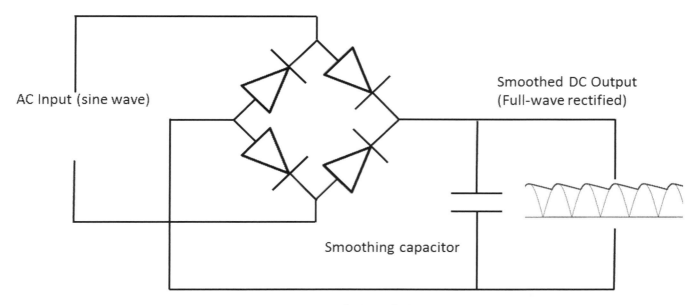

Figure 274 - A smoothed, rectified output

AC rectified DC current can be smoothed out using a capacitor. This collects charge when the current surges and releases it when the current falls, producing a smoother direct current, but still with some ripples.

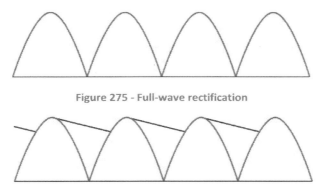

Figure 275 - Full-wave rectification

Figure 276 - Smoothed output using a capacitor

Transistors

One of the most important developments of the modern era was that of transistors. These allowed the development of computers and many other devices, including amplifiers and automatic control systems.

- They are used to amplify small amounts of current in one circuit to large amounts in another.
- Most known as NPN transistors, although there are also PNP transistors

A transistor has three pins, known as terminals: **Emitter, Base and Collector**

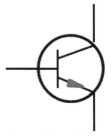

Figure 277 – NPN Transistor circuit symbol

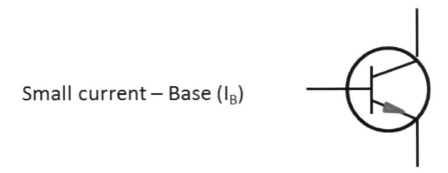

Figure 278 – An NPN transistor

This can be used to drive a large current through the transistor, using just a small current at the base the transistor will combine this current with that of the collector. The current in the emitter will be the sum of the collector and the base.

When there is no, or a very low, potential difference between the Base pin and the emitter, no current will flow between those pins, and no current will flow between the collector and emitter. The transistor acts as a switch, turning on a current through the collector and out of the emitter when the potential difference between the base and emitter reaches a certain level. At that point current flows both from a small base source and is added to that of the collector source.

The ratio of the collector current to the base current is called the **current gain**.

Quantity	Name	Units
Current Gain	Measure of the increase of current	Ratio
I_C	Current from collector	Amps (A)
I_B	Current through secondary coil	Amps (A)

$$Current\ Gain = \frac{Collector\ Current}{Base\ Current} = \frac{I_c}{I_B}$$

Some circuit components need low current to work, whereas others need larger amounts.

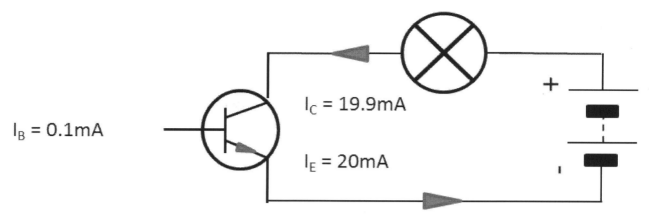

Figure 279 - In this circuit the 19.9mA is added to the 0.1mA to form a 20mA current

Using the gain equation above, we find that the gain of this circuit is:

$$\frac{19.9}{0.1} = 199$$

Here, the transistor has changed a SMALL AMOUNT of BASE CURRENT to a LARGER EMITTER CURRENT.

+Note that the conventional current direction must be in the same direction as the transistor arrow.

Practical Example – The Moisture Detector.

When a pair of moisture probes have a fluid between them, either because there is a liquid, or because there is moist air, a current will flow between the probed. The following circuit uses this to light up a bulb by causing a current out of the emitter and thus having a current going through the collector.

Figure 280 - Some transistors

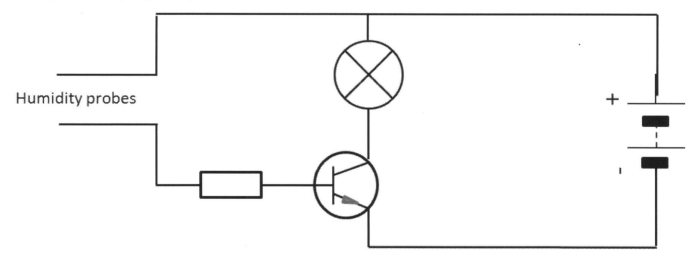

Figure 281 - When a current flows between the probes, a current will flow through the collector too, lighting the bulb

43.3 Questions

1) In a transistor, if the base current is 0.05mA and the collector current is 4.95mA, calculate

a. The emitter current
b. The current gain.
2) Define:
   a. Rectification.
   b. Half-Wave Rectification.
   c. Full-Wave Rectification.
   d. Smoothed Full-wave rectification
3) Draw a circuit diagram to show 3 bulbs in parallel and diodes in each of the circuits, all forward-biased. How many bulbs will light?
4) Name the following:
   a. An input device which detects changes in light.
   b. An output device which amplifies signals.
   c. An input device which stores electric charge.
5) Why is a capacitor often used in a rectifier circuit?
6) Analyse the following circuits:

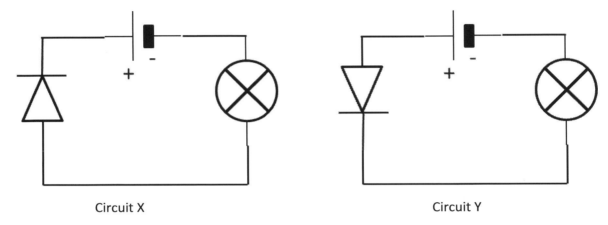

Circuit X   Circuit Y

a) In which circuit does the bulb light up?
b) In which circuit does the diode have a high resistance?
c) In which circuit is the diode forward-biased?
d) Draw on the conventional current direction.

7) What do the letters E, B and C stand for in a transistor?
8) Use the formula $I_E = I_B + I_C$ to calculate I_B if I_E = 27mA and I_C = 25.2mA.
   a. What is its bias?
9) What is the purpose of a rectifier?

Light-dependent resistor (LDR)
These components are often used in potential divider circuits where it is useful to switch on and off devices depending on the intensity of the light, e.g. **street lights, photography and burglar alarms**.

- Resistance **decreases** when light is **bright**.
- In the dark the resistance **increases**, and so the current drops

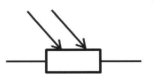

Figure 282 - LDR Circuit symbol

- Also known as a **photo-conductive cell**

Combining our knowledge of transistors and that of the varying resistance of an LDR depending on the level of light, it is possible to construct a simple, automatic, on/off switch for a street light. When the daylight is bright, the resistance of the LDR sensor will be low, and so the potential across it will be low.

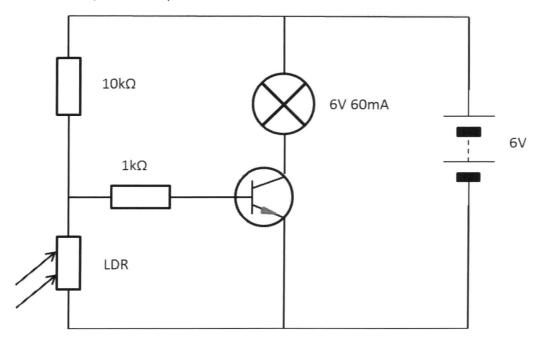

Figure 283 - Simple potential divider that can be used to turn on a street lamp when it goes dark

1) The LDR part of a potential divider circuit where the battery EMF is shared between the LDR and the fixed resistor.
2) The 1kΩ resistor placed in the circuit to prevent too much base current flowing into the transistor through the base and damaging it.

How does it work?

In the daylight the LDR has a low resistance and so the potential across it is low. This means the current will flow through it with ease and the potential across the base and emitter is also low. As a result the transistor stays **off**.

In the dark, the LDR increases in resistance, the potential difference across it, and therefore across the base and emitter of the transistor becomes high. As a result the transistor turns **on** and current passes through the collector, and therefore the bulb.

Capacitor for a time-delay switch

There are times when circumstances change rapidly but that it is not important to respond to immediately. For instance if a cloud passes over a street light, it might not make sense for the street light to turn on immediately. It might be better to wait a while before the lamp is turned on. Similarly, a burglar alarm might be too easily triggered if it went off each time a slight heat signature passed by. In both cases it is useful to add a time-delay switch.

- Used in street light (needs to be dark for a certain length of time before lights come on and burglar alarms.).
- Uses a **capacitor** in the circuit.

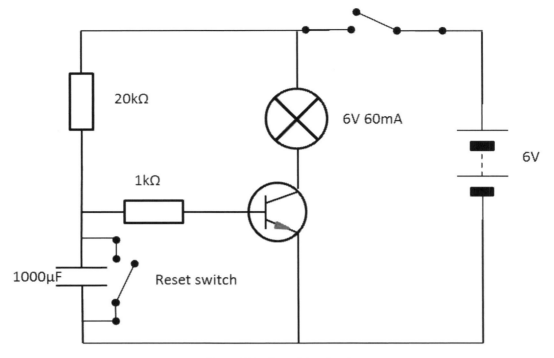

Figure 284 - Time delay circuit

How does it work?
1) Main switch closed but there is a time-delay before the bulb lights up.
2) This delay is due to capacitor in the potential divider circuit
    a. Capacitor charges slowly (the 20kΩ resistor slows down the charge flowing into the capacitor)
    b. Changing the resistor or/and capacitor will change how long it takes for the capacitor to charge up and thus how long the time-delay is for
3) Takes several seconds before voltage is high enough to switch transistor on.

The thermistor
This component can be used in a circuit that needs to respond to temperature changes, e.g. turning on and off alarms when it is too hot, or a cooler (or heater) when a certain temperature has been reached.

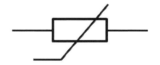

Figure 285 - Thermistor circuit symbol

The thermistor decreases in its resistance as the temperature rises, so the potential across it decreases as it gets hotter.

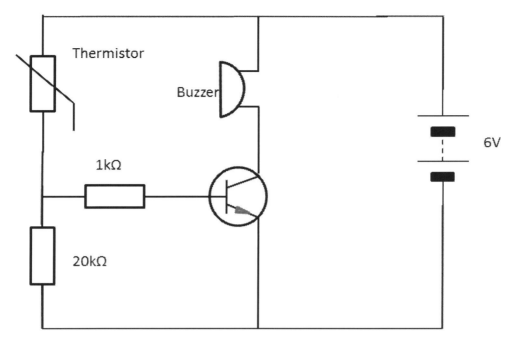

Figure 286 - An alarm that sets of a buzzer when the temperature is too high

How does it work?

1) Thermistor part of a potential divider:
   a. At room temperature, thermistor has **high resistance**.
   b. Also has most of the EMF.
2) **Lower resistor** does not have enough voltage to turn on the transistor.
3) When heated, the THERMISTOR'S RESISTANCE FALLS
   a. Lower resistor now has enough voltage to turn on transistor and ring the buzzer.

The light-emitting diode (LED)

These diodes emit light when a current flows through them and are used as indicator lamps, e.g. in TVs, Hi-Fi remote controls, burglar alarms and network hubs.

LEDs are sensitive and are not capable of reliably handling more than a 2V (4V in the case of a blue LED) potential difference. As a result, most circuits using these ensure that they are part of a potential divider circuit, with a fixed resistor acting to pull down the potential available to the LED.

Figure 287 - LED Circuit symbol

A typical diode would also allow only a 20mA current through it before it started to fail. Using Ohm's law it is possible to work out what the missing resistance is in the circuit below:

1) The potential allowed across the diode is 2V, so the resistor must drop the voltage from 6V to 2V, a potential difference of 4V
2) The current allowed in the diode is 20mA, so the current through the resistance is 20mA, or 0.02A
3) Using Ohm's law, R=V/I, we find that the resistance of the pull down resistor must be 2/0.02 = 100Ω

The circuit performs the same function as the circuit in Figure 286, except only an LED that is protected from the excess EMF is lit when the temperature rises too high.

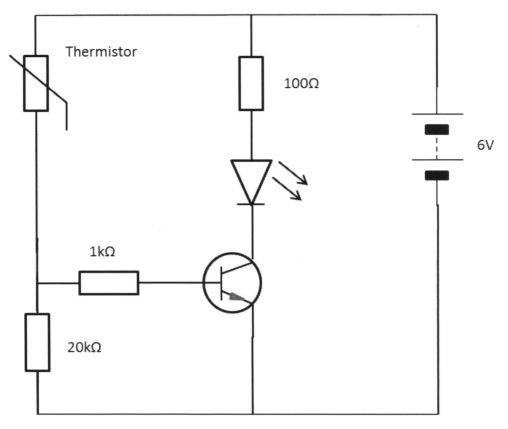

Figure 288 - this circuit turns on an LED when the temperature rises to a certain level.

43.4 Questions

1) Analyse the circuit diagram below. It is a temperature sensor. The temperature of the thermistor rises.

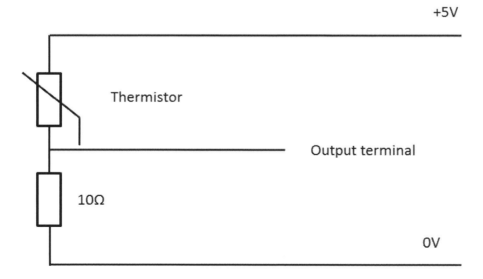

a. What happens to the resistance of the thermistor?
b. What happens to the voltage across resistor R?
c. Calculate the power dissipation (in kw) of resistor R if the current through it is 13A.

2) A pupil sets up the circuit below:
   a. Name F – J.
   b. Describe how the circuit works.
   c. Calculate the emitter current.
   d. Calculate the current gain.

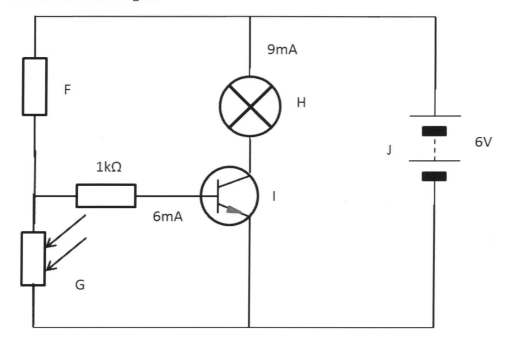

Chapter 44. Digital electronics

In the modern era there is a move to digitise all communications. There are benefits to this, primarily they are easier to correct when transmitted, and errors have crept in due to interference. As a result, **digital** signals can be clearer.

The alternative to digital transmission of information is using an **analogue** signal.

44.1 Analogue

Signals that are transmitted using analogue signals are simply **continuous** changes in voltages, often seen as sine waves.

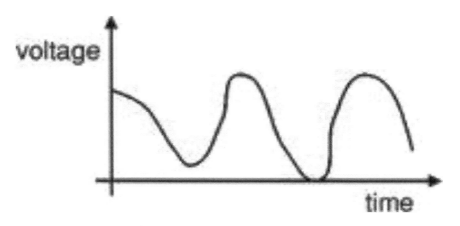

Figure 289 - Example of an analogue signal

All signals lose power during transmission; a process called **attenuation**. The loss of power makes the signal less distinct and may start merging with background interference, called **electrical interference**.

44.2 Digital

Signals that utilise digital transmission methods are represented by a either being a high value, called '1' or a low value, called '0'. This two value system, **either a 1 or a 0**, is called a **binary** coded system. The signal is thus sent as a series of pulses of known lengths. Unlike analogue signals, digital signals can be corrected at the receiving end, or during a relay amplification process.

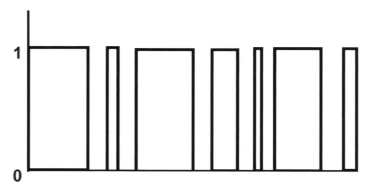

Figure 290 - example of a digital signal

Note how the signal only has two values, either 0 or 1. The voltage equivalent to the value "1" depends on the circuit, but is often in the region of 5V.

Chapter 45. Logic gates

All computers are based on making decisions based on sensors, or inputs and most devices that we use on a daily basis contain one, or many, computers, e.g. TVs, cars, buses, airplanes, mobile phones, washing machines, burglar alarms and so on.

The fundamental building blocks of all computers are **logic gates**, which are best considered as simple electronic switches. All logic gates work as follows:

1) Take 1 (or more) binary input values, i.e. 0s or 1s
2) Perform some kind of comparison between the values
3) Output a 0 or 1 value

Step 2 above depends on the logical comparison that is being made, or, as in the case of the NOT gate (see later), it simply changes a 1 to a 0, and vice versa.

45.1 Truth tables

In the following sections we will be making extensive use of truth tables to illustrate what occurs when certain input values are passed through various steps for comparison.

They are very simple to understand. In the following table, the first two columns represent the values that are to be compared with each other. In this table we call these values A and B, and the result we call O. The value A can take on the values 0 or 1, and similarly, B can be either 0 or 1 too. In the first two columns we write down all the possible combinations of A and B, making sure that there are no combinations repeated. For two inputs this will always be 4 combinations. For this truth table the output, O, will answer the question if both A AND B are the value 1. For O to take on the value 1, both A AND B must be 1, which is only true in the last row.

A	B	O
0	1	0
0	0	0
1	1	0
1	0	1

This truth table is the one used for an AND truth table. We will see more examples later, e.g. an OR table.

45.2 AND logic

In the following circuit we will notice that for the lamp to light up, both switches A AND B must be closed. This is a simple version of an AND circuit and might be used in a washing machine to ensure that the door, which operates a switch when it is closed, thus ensuring that both the on/off switch and the door has to be closed for the machine to operate.

Switch A	Switch B	Bulb
Open	Open	Dark
Closed	Open	Dark
Open	Closed	Dark
Closed	Closed	Lit

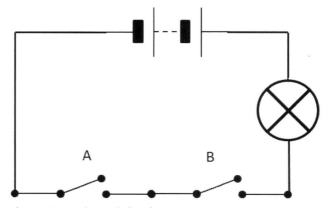

Figure 291 - Using switches for an AND circuit

This kind of logic gate can also be created using transistors, one connected in series with the other, as seen in the circuit diagram below. The input voltages at both A and B must be high in order for the emitter in the lower transistor to provide a current to the bulb. The same truth table will apply to this as for the switch circuit.

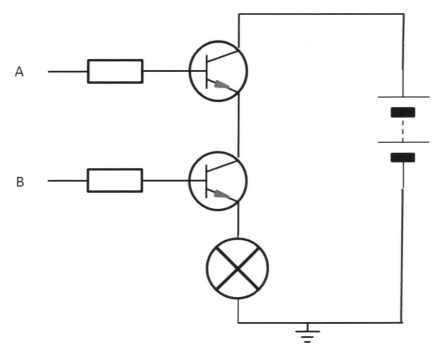

Figure 292 - Using transistors to provide an AND circuit

The two circuits above are examples of logical AND gates and are used extensively in electronic circuits. When drawing this as a part of a logic diagram there is a standard symbol that is used.

The standard truth table for this is seen to the left.

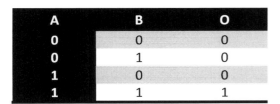

Figure 293 - AND symbol

A	B	O
0	0	0
0	1	0
1	0	0
1	1	1

45.3 OR Logic

In the following circuit, either switch A OR switch B can be closed for the bulb to light up. The logic table for this situation shows that either A or B can be closed for the desired output.

Switch A	Switch B	Bulb
Open	Open	Dark
Closed	Open	Lit
Open	Closed	Lit
Closed	Closed	Lit

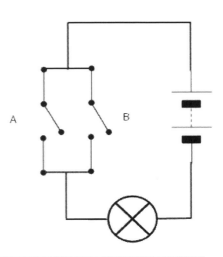

Either A OR B can be closed for the bulb to light up. It is worth noting that this also includes the situation where both A AND B are closed. If one were to only want to switch it on when either A or B are switched on, but not both, an exclusive OR gate would be used. These are known as XOR gates and are covered later.

The logical version of this truth table is seen to the right.

In common with the AND gate above, the OR circuit can also be created out of transistors. In the following circuit diagram the emitter of either the top OR the bottom will provide the current to the bulb when their bases are at a high potential. This could have a use in a car where the inside light will go on when either the driver or the passenger door is opened, or both.

A	B	O
0	0	0
0	1	1
1	0	1
1	1	1

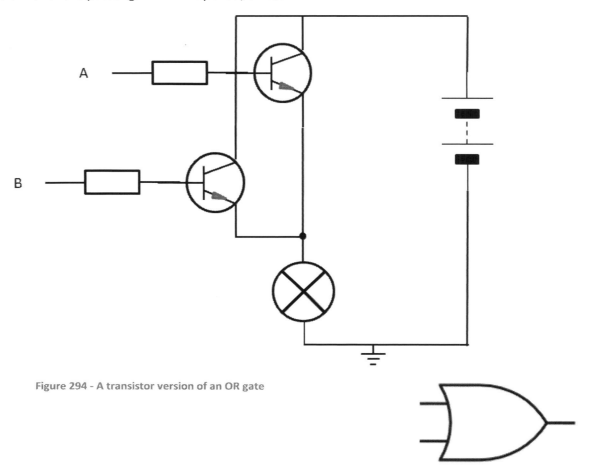

Figure 294 - A transistor version of an OR gate

The logic diagram circuit symbol for an OR gate is depicted to the right.

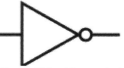

Figure 295 - OR Symbol

45.4 NOT Gates (called an "inverter")

NOT gates simply change a single input to the opposite on the way out.

I	O
0	1
1	0

Quite often these are seen in places where something needs to be turned on when something is NOT true, e.g. a street light

Figure 296 - NOT symbol

comes on when it is NOT light, an oven is turned on when the food is NOT cooked.

Note the circle at the end of the NOT symbol. This circle, when combined with OR and AND gates will invert their output, as seen below.

45.5 NAND Gates

This is a combination of an AND gate and a NOT gate. The long version of drawing this logic circuit is shown below, but, since this is a common combination of logic circuits, a symbol exists that combines the two:

Figure 297 - the long version of a NAND gate

Figure 298 - Logical NAND gate - notice the circle

A	B	A AND B	O
0	0	0	1
0	1	0	1
1	0	0	1
1	1	1	0

45.6 NOR Gates

As in the NAND gate, the NOR gate combines two gates and has a long and short version of its logical symbol

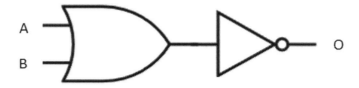

Figure 299 - long version of a NOR gate

Figure 300 - Logical NOR gate

A	B	A OR B	O
0	0	0	1
0	1	1	0
1	0	1	0
1	1	1	0

45.7 Combining logic gates

We mentioned, above, that the OR gate included both the OR and the AND logic. It is possible to create a simple combination of logic that removes the AND part of that comparison. Its truth table would be:

A	B	O
0	0	0
0	1	1
1	0	1
1	1	0

We will not go through the derivation of the combination of logic to perform this, but it can be written as:

$$(A\ AND\ NOT\ B) OR (B\ AND\ NOT\ A)$$

Using our normal symbols to build this, we will end up with this logical circuit:

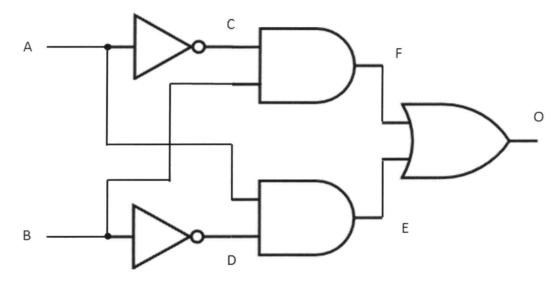

When analysing this kind of circuit it is best to do so step by step. In the above diagram the inputs are A and B, the final output is O, but there are several steps before the result is seen, each one being given a letter, C through E. The first step to analyse it is to draw a truth table with each step in a column.

A	B	C (NOT A)	D (NOT B)	E (A AND D)	F (B AND C)	O (F OR E)
0	0	1	1	0	0	0
0	1	1	0	0	1	1
1	0	0	1	1	0	1
1	1	0	0	0	0	0

45.8 Questions

1) Examine the circuit to the right
   a. Write a truth table for the diagram
   b. What type of logic gate is this?

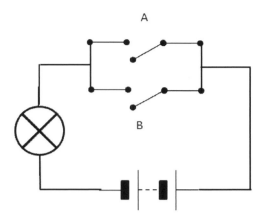

2) The diagram below shows two logic gates in series. Write a truth table for this combination.

3) Write a truth table for the diagram below:

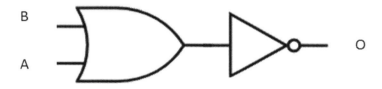

4) The diagram shows an alarm system. When the light sensor detects light it is true (or 1), and when the door sensor senses that the door is open it is true.

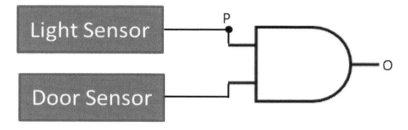

   a. When will the alarm sound (O becomes true, or 1)?
   b. A NOT gate is placed at point P. what effect will this have on:
      i. The input to the AND gate?
      ii. The conditions under which the alarm will sound?

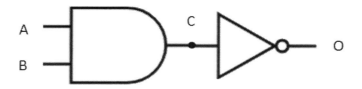

5) Complete the truth table for the above system.

A	B	C	O
0	0		
0	1		
1	0		
1	1		

Chapter 46. Cathode ray oscilloscopes (CRO)

Cathode ray oscilloscopes are devices that trace an input based on the value of its voltage across a screen. It does so by guiding a stream (or ray) of electrons onto a fluorescent screen, where the ray can be seen as a glowing dot.

The ray is moved across the screen from left to right at a user-selected rate so that the resulting dots produce a continuous line, showing how the input voltage is changing over time.

The rays are called cathode rays because it consists of a stream ofe negatively charged particles, called electrons. Compare this with your understanding of **cathode** from chemistry. These electrons are produced by heating wires, similar to a tungsten light filament,

which then releases electrons. This is called **thermionic emission**.

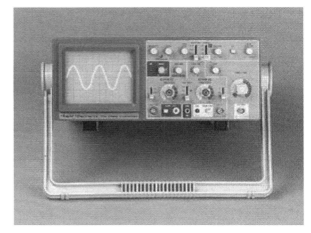

Figure 301 - CRO with a trace on its screen

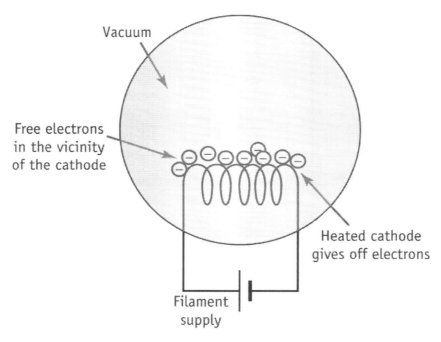

Figure 302 - Thermionic Emission - electrons "boil off" a filament

In Figure 302, the electrons simply exist around the filament but do not stream. To do this we need to accelerate them. This is done by attracting them to a positively charged plate. By placing a hole in the centre of this plate, the electrons will accelerate towards it and then pass through the hole, forming a ray of electrons, a cathode ray.

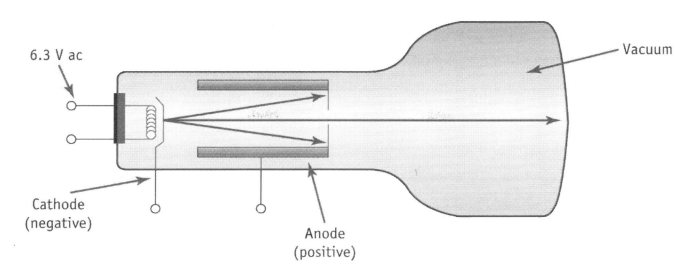

Figure 303 - The anode attract the thermionically emitted electrons, causing a stream of them through a hole towards the right

At the end of the vacuum tube, on its inner surface, is a fluorescent screen. When the electrons hit this spot it glows.

The ray is deflected up and down depending on the voltage of the input signal. The higher the voltage, the more the ray is bent upwards, and the lower the voltage (or if it is negative) the lower it is guided. This is down by have two horizontal plates (called **Y-plates**) that the ray passes through.

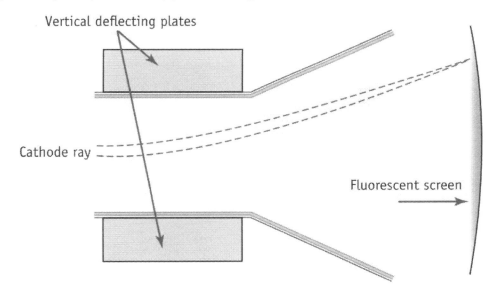

Figure 304 - The ray is deflected up or down depending on the input voltage

The last step in controlling the ray is to make it move from one side to the other and then continue doing this. This is done by adding two vertical plates (called **X-plates**) where the potential difference between the plates is changed at a user-selected rate, causing the ray to deflect from left to right, and then starting again from left.

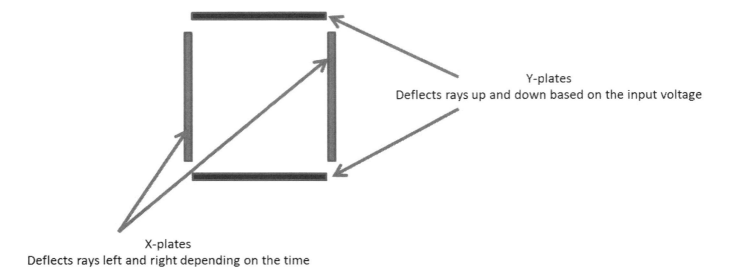

Figure 305 - Deflector plates attract and repel cathode ray

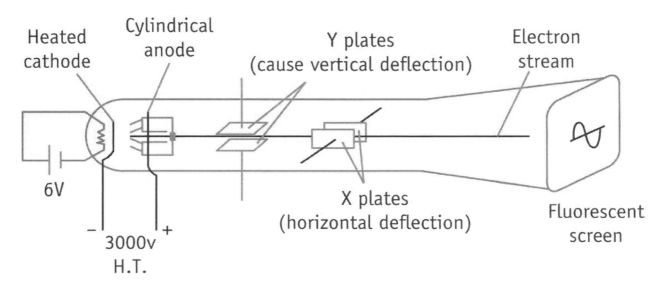

Figure 306 - The complete assembly as used in an OCR. The X-plates cause a deflection in the horizontal plane and the Y-plates in the vertical

The above system works by causing the deflection through attraction (and repulsion) because of the charge of the ray, and that of the X- and Y-plates. It is also possible to cause this deflection using a magnetic field since a cathode ray is simply just a flow of charge. To determine the direction of the ray's deflection, use Fleming's left hand rule, but remember that the current is the way a positive charge flows, the conventional current. So you must point the I in the opposite direction to the cathode ray.

46.1 The CRO Display and controls

At first sight the CRO can seem very intimidating. We will not cover every part of the following diagram, instead focusing on the main controls and display.

The display

The display is divided up into a grid. Each division in the X and Y axis either measures the value of the input voltage (the Y-axis) or how long (the X-axis). The size of each division is set using the controls.

The ray will move at across the screen at the rate given by the **time/div** knob at the top of the diagram below. The longer the time set, the slower the ray will move across the screen. By using this information, and measuring the number of divisions, it possible to work out how long a particular change has taken. In the case of a simple waveform, it is possible to work out the period, and this the frequency.

The time/div and volts/div adjustments are often called **gains**.

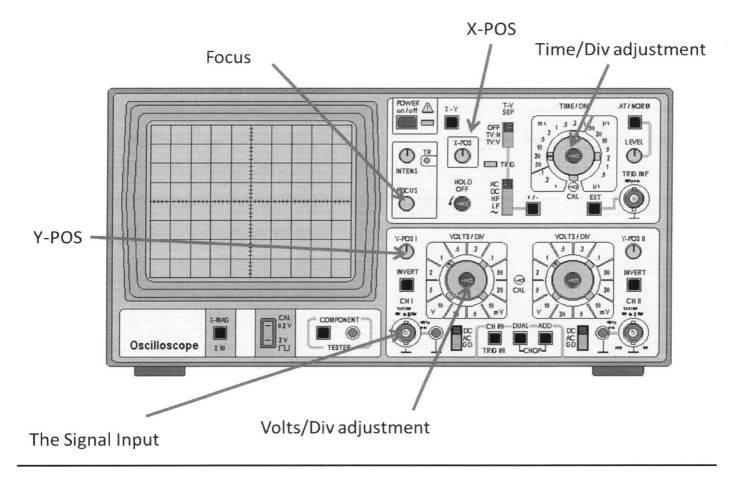

Figure 307 - A typical oscilloscope

The **Volt/Div** adjustment changes how much each division represents in terms of the voltage it is displaying. A larger value here means that each division represents a larger potential in the input signal.

The **Focus** knob allows the user to make the ray more distinct as it hits the screen. This can help in making more accurate measurements.

In addition, there is a **X-POS** and **Y-POS** know that allows the user to "zero" the ray initially.

46.2 Studying waveforms

In the following screen shots from a CRO it is possible to see various traces that illustrate many of the topics we have covered in the chapters above.

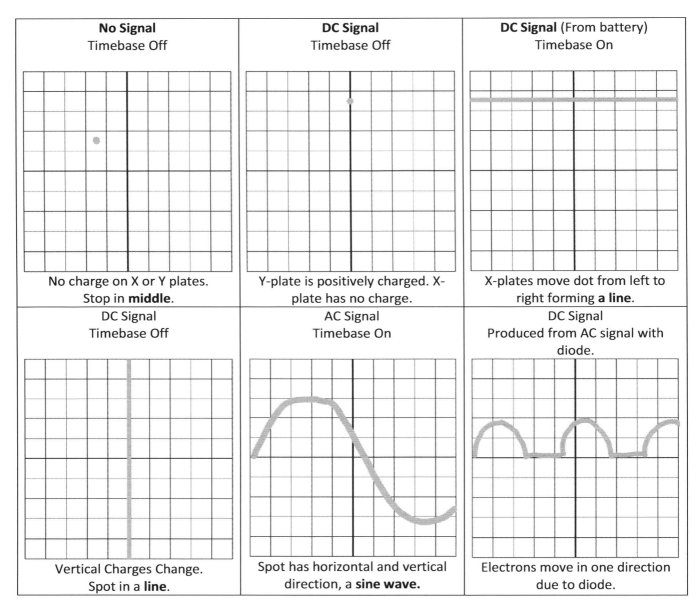

46.3 Measuring voltage

It is now possible to measure a number of items directly from the OCR screen by counting the divisions. In the following screen, the Time/Div is set to 2s, and the volts/Div is set to 5V.

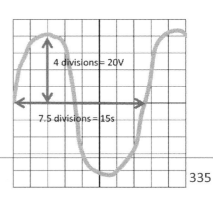

The amplitude is 4 divisions; each division is set to be 5V, so the peak voltage is 20V. The wave length is 7.5 divisions, each division is set to be 2s, so the period of the wave is 15s. Its frequency is 1/period, or 1/15=0.07Hz.

46.4 Questions

1) The diagram shows a cathode ray oscilloscope

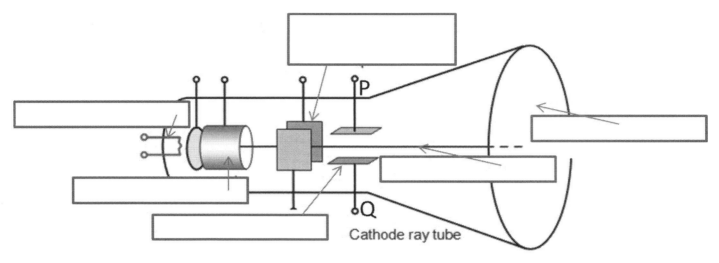

   a. Label the parts on the diagram.
   b. Explain how the filament is made to produce electrons.
   c. P and Q are deflection plates. Complete the path of the electrons.
2) Define the term "Thermionic Emission."
3) The amplitude of a wave on an oscilloscope is 2.4cm. Calculate the peak voltage if the gain is set to 12V/div.
4) Label the +ve and –ve charges on the plates below based on the location of the orange labelled cathode ray

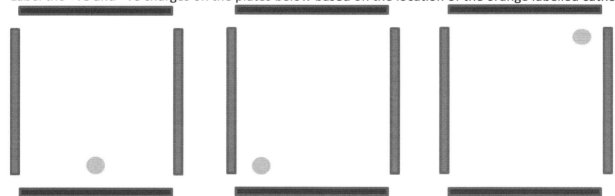

Chapter 47. Atomic Physics

47.1 The Atom

- The word atom is from Greek and means "indivisible" – something that cannot be divided further
- Although this original meaning has since been shown to be incomplete, an atom is the smallest particle that can take part in a chemical reaction

Atomic Model

- Ernest Rutherford, performed the most important experiment to explore what atoms were made out of

Figure 308 - Ernest Rutherford

Rutherford's Alpha Particle Experiment
We will cover what alpha particles are in more detail later. For now think of them as small, highly, positively, charged, particles; larger than electrons, but smaller than gold atoms.

Rutherford fired alpha particles at a very thin gold film. Surrounding the gold film was a circular enclosure where the walls were covered with a special coating that showed when an alpha particle hit it. The prevalent model of the atom at the time was called the "plum pudding" model. This model, proposed by the discoverer of the electron, J.J. Thomson, had the negative charges of the electron mixed in with a positive charge, with the electrons being stuck on the positive charge, like plums in a plum pudding. In his theory, proposed in 1906, the atom was thought to be a sphere, one billionth of a metre across.

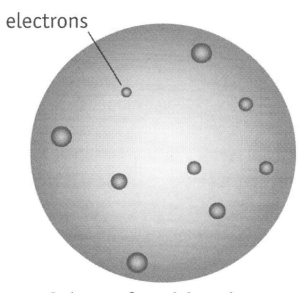

Figure 309 - Plum pudding model with electrons baked into a positively charged atom

- If the plum pudding model was correct, the highly charged alpha particles should pass though the gold film, maybe deflecting a little bit to one side or the other.
- The two experimental researchers (Geiger and Marsden), however, noticed that this was far from true and 1 in every 8000 alpha particles bounced back, whereas most passed through with some deviation.

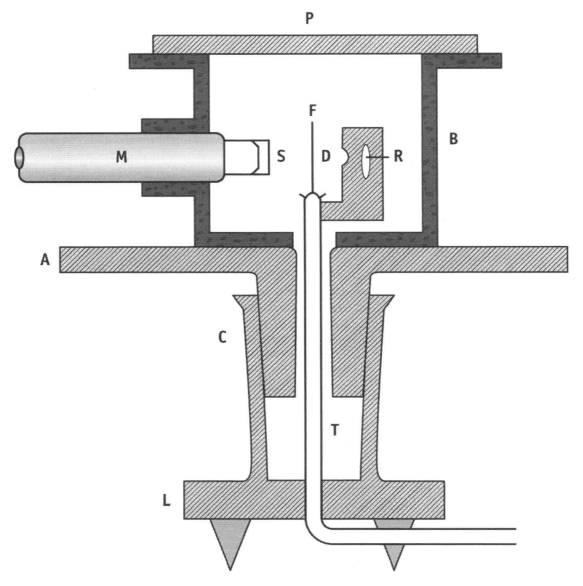

Figure 310 - Geiger and Marsden's apparatus. Alpha particles (from R) scatter off the foil (F). The microscope (M) is rotated around the cylindrical box (B) to count scattering at any angle

The conclusion was essential to the understanding of how atoms were made up.

- Most of an atom is empty space.
- There is a solid nucleus to the centre of the atom

- Since the alpha particles are positive, and positive particles repel positive particles, the nucleus had to be positive.
- Electrons were seen as orbiting the nucleus, like planets around a star

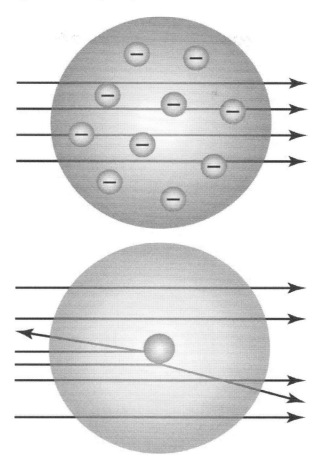

Figure 311 - Deflections of alpha particles that hit a "plum pudding" atom at the top, and what they found, at the bottom

In an iGCSE you might be asked to show how the alpha particles scattered in this experiment. The diagram below shows the types of paths that alpha particles would take depending on where they hit the gold particle.

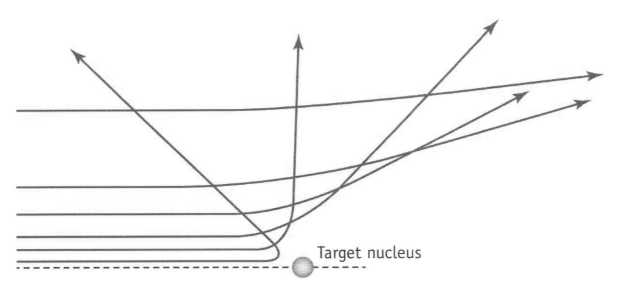

Figure 312 - The paths that the alpha particles were observed taking

Chapter 48. The Model of the atom

Following the experiment with alpha scattering it was found that the nucleus of the atom consisted of two distinct particles, and so the complete picture of the atom is:

- A nucleus that consists of:
    - A proton – a positively charged particle with the same strength of charge as the electron (see below), but opposite.
    - A neutron – a neutral particle and has nearly the same mass as a proton
- Negatively charged electrons that orbit the nucleus at high speeds and have nearly no mass

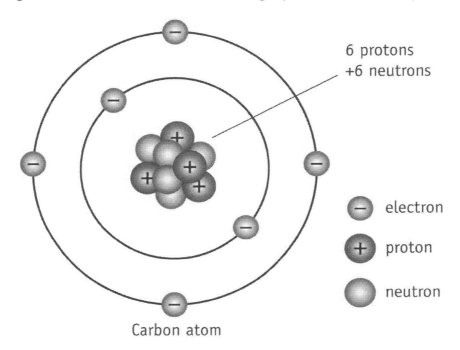

Figure 313- Atomic structure, as seen in a carbon atom

In the carbon atom, seen above in Figure 6, it can be seen that the nucleus has 6 protons and 6 neutrons[1].

- The number of protons is the same as the element number in the periodic table (see next page, where carbon is circled)

The element therefore changes if you add a proton to the nucleus. It is, however, possible to add neutrons to a nucleus and not change the element. When the same element has several possible configurations of neutrons, they are called isotopes.

> An Isotope is "Each of two or more forms of the same element that contain equal numbers of protons but different numbers of neutrons in their nuclei"

- An isotope is "Each of two or more forms of the same element that contain equal numbers of protons but different numbers

---

[1] This image is a little misleading though, as the distance between the nucleus and the electrons is massive, when explored at the atomic scale.

of neutrons in their nuclei"
- Isotopes are chemically identical

For Carbon there are several isotopes, the most common one having the configuration seen above, with 6 protons and 6 neutrons. One other, very important isotope has 8 neutrons and is called Carbon 14 (see below). This is still carbon as it has 6 protons, but is unstable and will break down (decay) in time. As we will see later, Carbon 14 is a vital tool for dating organic material.

Nomenclature

As with chemistry, physics has a standard method of writing the information about isotopes:

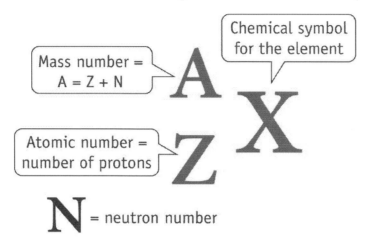

Figure 314 - how to write down a nuclide

- A - Mass number (or Nucleon Number) – the total number of protons and neutrons in the nucleus (For carbon 14 this is 14)
- Z – Atomic number – the number of protons. For carbon 14 this is 6
- X – The Chemical symbol for the element. For carbon 14 this is C
- Nuclide – an isotope with a known Mass number and Atomic Number

The common forms of carbon are Carbon 12, Carbon 13 and Carbon 14. These can be written down in the forms seen to the right, using the method shown above.

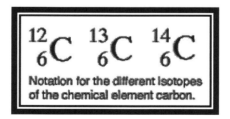

Notation for the different isotopes of the chemical element carbon.

Follow the steps below to create the most common version of Helium and note how the circled symbol changes while you add electrons, protons and neutrons.

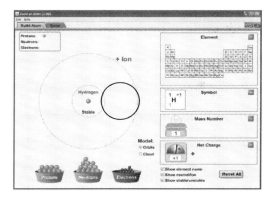

Figure 315 - Building an Atom - Hydrogen without an electron, as an Ion

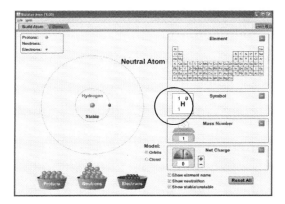

Figure 316 - Adding an electron balances the proton charge with the electrons' opposite charge

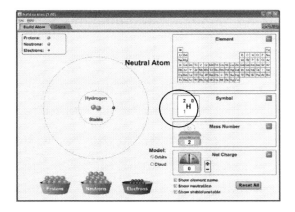

Figure 317 - Adding a Neutron to the nucleus doesn't add charge, but does create a an isotope of Hydrogen

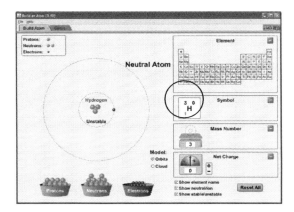

Figure 318 - adding another Neutron still keeps it as Hydrogen, but with two Neutrons. Another isotope of Hydrogen

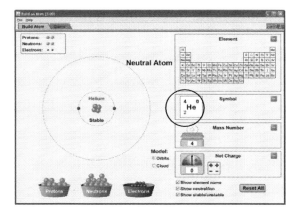

Figure 319 - adding another proton changes the Hydrogen element to Helium. Adding an electron ensures it is not charged

Questions

1) Complete the missing information in the table below

Element	Atomic No.	Mass No.	Protons	Neutrons	Electrons
Aluminium	13			14	
Carbon	6			7	
Neon			10	12	
Chlorine		37	17		
Gallium				40	31
Potassium	19			22	
Titanium		48			22
Cobalt		59			27
Chlorine	17	35			
Copper	29			34	
Cobalt		60			27
Gallium	31	69			

Chapter 49. Radioactivity

49.1 Background Radiation

There are many sources of background radiation; the most common ones are:

- Unstable isotopes around us
    - Radon (often released from Granite rocks)
    - Remnants of past nuclear explosions
    - Radioactive rocks
- Space
    - From the sun and other stars

The background radiation is constant and, usually, not intense, and therefore not usually dangerous.

49.2 Activity

One of the ways of determining how radioactive a substance is is to aim a GM counter at the sample and see how many decays there are each second. This is measure in becquerel. One becquerel (Bq) means that one decay occurred that second, 2µBq means that 2×10^6 decays took place that second.

- Activity – the number of decays occurring in a period of time.
- Measured in becquerels (Bq)

49.3 Forms of Radioactivity

When an isotope is unstable it will wish to decay through one or more methods. The particles that are released in these decays are called radioactive particles. There are three forms of radioactivity:

- Alpha (α) Particles – these are made up of two protons and two neutrons, but no electrons
- Beta (β) particles – these are high energy electrons that are released from the nucleus of an atom (see more on this below)
- Gamma (γ) rays (not particles) – these are very high energy and high frequency electromagnetic rays that are emitted from the nucleus of a nuclide

The different forms of radioactivity can be explored from several aspects and the iGCSE papers typical explore the following ones:

1) Penetration – what material can the radiation penetrate
2) Hazards – what kind of danger is it
3) Protection – How can we protect against the radiation

49.4 Alpha (α) Particles

These particles are very large, compared to the other two forms of radiation, and has a POSITIVE charge of +2, because it has two protons and no electrons to provide a neutralising effect. As such they are very likely to interact with other particles that they pass by. They do this by attracting electrons and thus making it neutral again (and a Helium element). As a result it is highly ionising (causing ions to be formed) and damages organic tissue and can cause cancers. It also will not travel very far before attracting the electrons, typically no more than 10cm in air, and cannot penetrate very far, typically being blocked by as little as a piece of paper.

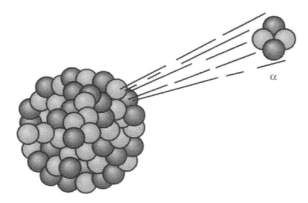

Figure 320 - Alpha particle decay from a larger atom

Alpha particles are formed when heavy isotopes go through fission (when the split apart).

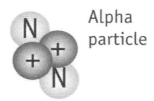

Figure 321 - Alpha particle, two protons (+) and two neutrons (N)

Symbol for an Alpha particle is: $^{4}_{2}\alpha$.

49.5 Beta (β) particles

These high energy electrons are very small and move very quickly. They are normal electrons in all ways except for that they emerge from a nucleus when a neutron splits into a proton and the beta particle. As such, the nuclide loses a neutron, but gains a new proton and thus becomes a new element. We will explore this more later.

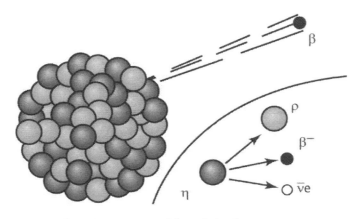

Figure 322 - Beta particle emission from an atom

As a high energy electron it can either dislodge another electron orbiting an element, or join as an extra electron (less likely). As such it can ionise other particles and cause damage to living tissue, but not as much as alpha particles. Beta particles are fairly easily absorbed by a thin sheet of aluminium or over a few metres of going through air.

Symbol for a Beta particle is: $_{-1}^{0}\beta$

49.6 Gamma (γ) rays (not particles)

When the nucleus of a nuclide has surplus energy it becomes unstable and can emit a gamma ray. These rays (they are not denoted as particles in iGCSE) have very high frequencies, and therefore energy. You will have learned about these in the Waves section of the iGCSE syllabus. Since they have very high energy, they are capable of exciting one more electrons such that these leave the orbit of their original elements, thus leaving an ion behind. As such the radiation is damaging to organic tissue, but not as damaging as either beta or alpha particles. Gamma radiation is, however, very penetrative and it is not possible to completely protect from it, although thick lead will absorb much of it. Gamma rays will therefore travel forever and are seen emanating from distant galaxies. Since gamma rays are not particles, they are not charged.

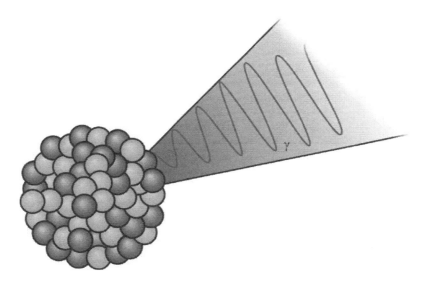

Figure 323 - Gamma ray emission from the nucleus of an atom

49.7 Summary

The table below shows the above information in summary form.

Table 14 - summary of radioactivity

Radiation	Particle	Ionisation	Penetration	Travels in air	Charge
Alpha	Yes	Most harmful	Stopped by paper	10cm	+2
Beta	Yes	Harmful	Stopped by aluminum	Several metres	-1
Gamma	No	Least harmful	Hindered by lead	Forever	None

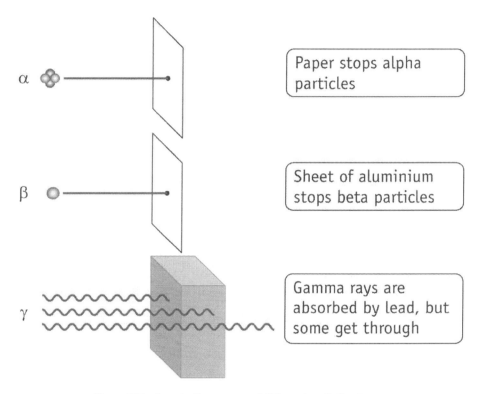

Figure 324 - Penetrative power of different radiation types

49.8 Deflection

As outlined in the sections and table above, alpha and beta particles are charged and as such will be deflected when passing between two charged plates, or through a magnetic field. The deflection in an electric field follows the simple rules of likes repelling and opposite attracting.

- Beta particles deflect towards the positive terminal
- Alpha particles deflect towards the negative terminal.
- Gamma particles have NO charge, so do not deflect

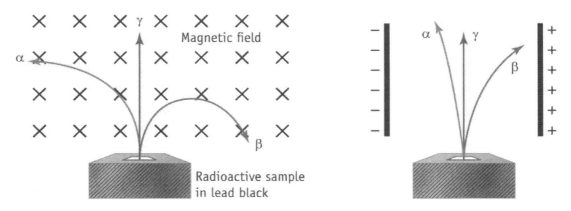

Figure 325 - How a magnetic field and charged plates affect the movement of a radioactive particle

Deflection due to a magnetic field uses Fleming's left had rule:

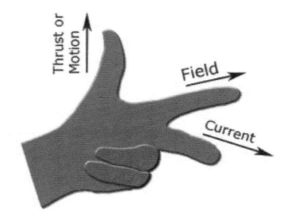

It is important then to remember that this rule uses the **conventional current definition** and so it is the direction a positive charge moves – as such alpha and beta particles will defect in opposite directions. In the following diagram from a past paper, you are asked to determine which decay goes in which direction. The one that goes straight through will thus be gamma, since it is not charged, the one going to the top left will be alpha particles since the deflect less than beta particles due to their heavy mass, and the one to the right is therefore beta particles.

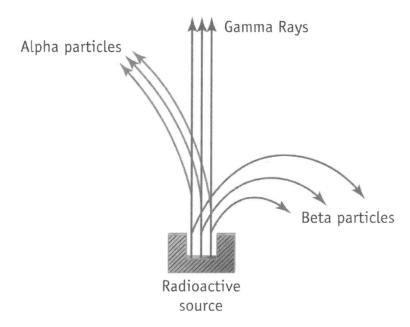

Figure 326 - Deflection paths of different particles passing through a magnetic field

49.9 Questions

1) The diagram below shows a beta particle passing between two magnets. Use Fleming's Left Hand rule to determine the movement of the particle.

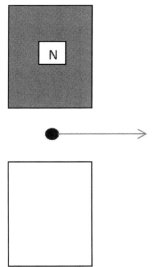

2) Complete the table below, using the units set out in the first example

Particle/Ray	Mass	Charge
Proton	1	+1
Neutron		
Electron		
Alpha particle		
Beta particle		
Gamma ray		

3) Draw diagrams showing how each radiation type (alpha, beta and gamma) are deflected when passing through an electrostatic field
4) The symbol:

$$^{35}_{17}Cl$$

represents one atom of chlorine.

   a. State the types of particles in this atom and how many of them there are
   b. State where each particle is to be found in the atom
5) Radiation from a radioactive sample was taken over a 6 year period. The table shows the results from the sample. The background radiation count was 4 per minute.

Time (years)	0	1	2	3	4	5	6
Activity count (counts per minute)	124	80	52	34	23	16	12
Activity due to the sample only							

   a. Complete the table
   b. Plot the values of the activity of the sample against time
   c. Join the values with a smooth curve

49.10 Ionisation

The word ionisation means to create an ion. This term is used in chemistry in the same way, when an element, or compound, has a net charge. This causes an ion to seek a new electron, possibly through a bond. As if this occurs in living tissue, such as within a DNA, it can cause cancerous growths in cells and is thus harmful.

- Ionisation causes ions, that can, in turn be harmful to living organisms

The most harmful of the three radiations is alpha!

Chapter 50. Detection of radiation

50.1 Photographic film

When photographic film is exposed, in darkness, to a radioactive source, it turns black. This can be used to determine if there is radioactivity in the area, cheaply and is used in industry in badges that are analysed at the end of a working day.

50.2 Gold leaf electroscope

When radiation passed by the gold leaf, it will ionise the air there. Since the air will be short of electrons, the surplus electrons in the gold leaf will leave the leaf and join the ions in the air instead. This will cause the leaf to have less charge and drop back down.

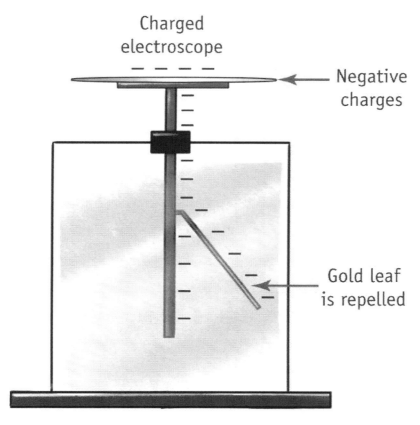

Figure 327 - Gold leaf electroscope

50.3 Geiger-Muller Tube

Geiger was one of Rutherford's students, who was mentioned above. He invented this tube so save himself time in observing the alpha particles. His invention allows radiation to enter through a fine "window" at one end of the tube into a low pressure gas (usually argon – Ar). When an ionising particle enters the tube and ionises the gas, it sets off a high voltage spark in the tube, and a current. This is amplified and usually made audible with a loudspeaker.

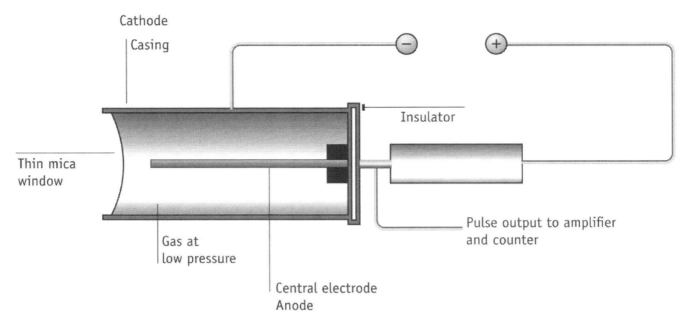

Figure 328 - GM Tube

50.4 Cloud Chamber

A chamber is filled with cold alcohol vapour and is illuminated with a light. When charged particles pass through the gas, it causes it to condense and leaves a path of tiny droplets showing its path through the chamber.

Figure 329 - Cloud Chamber

- Alpha particles make **Thick, straight, lines**
- Beta particles make **Thin, wavy, lines**
- Gamma particles make **Thin, straight, lines**

50.5 Spark Counter

This works best with alpha particles. A gauze and a wire are arranged as below and a potential difference it applied between the two that is just short of causing sparks to fly between them. When a charged particle passes through the gap, it will ionise the air and causes the resistance of the air to drop and a spark will result between the gauze and the wire.

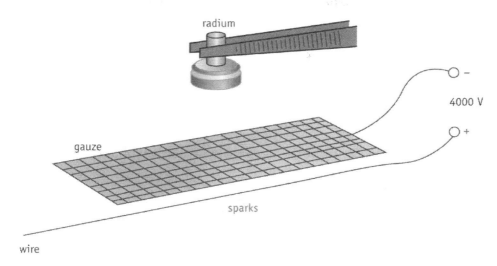

Figure 330 - Spark Counter

Chapter 51. Uses of radioactivity

Isotopes that are unstable (radioactive) are called radioisotopes. These can be used in various situations to help with problems:

51.1 Tracers

Small quantities of low radiation radioisotopes are added to drinks, or other parts of an organic system and can be detected as the move through the human body, or a plant. They can also be used to detect leaks in underground pipes.

51.2 Radio therapy

Cobalt 60 is a powerful source of gamma rays, and these can be used to kill off cancerous cells in the body

51.3 Thickness measurement (guaging)

In paper and iron mills, the thickness of the rolls can be controlled by measuring how much beta radiation passes through the sheet to a Geiger counter.

The counter controls the pressure of the rollers to give the correct thickness. With paper, or plastic, or aluminium foil, beta particles are used, because alpha will not go through the paper.

They choose a source with a long half-life so that it does not need to be replaced often.

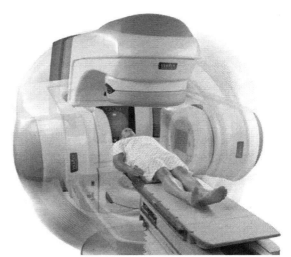

Figure 331 - Radio therapy

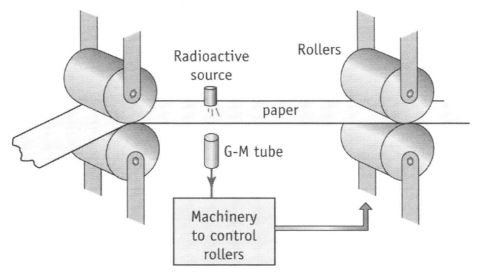

Figure 332 - Measuring the thickness of paper

51.4 Carbon 14 Dating

The amount of carbon 14 in all living beings stays constant while they are alive, but when they die this unstable isotope starts to decay (into nitrogen, as we shall see

When they die they stop taking Carbon in, then the amount of Carbon-14 goes down at a known rate (carbon 14 has a half-life of 5700 years). The age of the ancient organic materials can be found by measuring the amount of carbon 14 that is left.

51.5 Sterilising

Even after it has been packaged, gamma rays can be used to kill bacteria, mould and insects in food. This process prolongs the shelf-life of the food, but sometimes changes the taste. Gamma rays are also used to sterilise hospital equipment, especially plastic syringes that would be damaged if heated. This is called irradiation

Figure 333 - Irradiated operating equipment

51.6 Question

1) The table below gives the activity of an isotope over time:

Time: (mins)	0	10	20	30	40	50	60	70	80	90
Activity (counts/minute)	96	78	62	54	40	32	26	21	15	14

    a. Plot a graph to show how the activity of the sample changes with time
    b. Predict a count rate at 100 mins

2) A radioactive substance emits beta particles only. The count rate was taken every 20 minutes:

Counts/minute	330	231	165	120	90	71	57
Time (minutes)	0	20	40	60	80	100	120

A count rate of 30 was obtained even without the source nearby.

    a. What was the cause of the rate of 30 counts per minute?
    b. Draw a table to show how the count rate due to the beta-source varies with time
    c. Plot a graph of the beta count rate against time
    d. What effect does the emission of a beta particle have on an atom?

3) Name 3 uses of radioisotopes
4) Give two uses of gamma radiation
5) Why are gamma rays rather than beta or alpha particles used to kill cancer cells?

51.7 Safety precautions when handling radioactive materials

Radioactive sources which are used in school are usually very weak. They can only be used in the presence of an authorized teacher. They are kept in a sealed container except when they are being used in an experiment or demonstration. They are immediately returned to the container when the experiment or demonstration is finished.

Always display the radioactive warning/hazard symbol.

Figure 334 - Radiation Hazard Symbol

When using the radioactive source it should be

- Handled with tongs or forceps, never with bare hands.
- Kept at arm's length, pointing away from the body.
- Always kept as far as possible from the eyes.

Figure 335 - Wearing Radiation Suits

Hands must be washed after the experiment and definitely before eating. Radioactive sources which are used in industry may be very strong and additional precautions must be followed. Fully sealed protective suits can be used which prevent any direct contact with radioactive materials. The suit can be lead lined to prevent penetration by a-particles and b-particles. Robots which are operated by remote control can be used to deal with very contaminated areas or high intensity gamma-emitters.

People who work with weak radioactive sources wear a film badge.

Chapter 52. Radioactive Decay

All heavier elements are capable of being split into lighter elements through the emission of alpha and beta particles. Gamma rays are emitted from certain elements, but do not change them. The process of emissions of alpha and beta particles is called decay.

- The decay is random. Each element has a possibility of decaying immediately, at some random time, or never.
- When decays occur, new elements are formed
- Unstable elements usually decay into more stable elements

52.1 Decay balancing equations

As with chemical reactions, these equations must balance on each side. The sum of the mass number and the atomic number must be the same on the left and right. This is why the Beta particle has an atomic number of -1.

Alpha Decay
- An element loses two protons and two neutrons
- The mass number is reduced by 4, and the atomic number by 2

Examples:

$$^{238}_{92}U \rightarrow {}^{234}_{90}Th + {}^{4}_{2}\alpha + Energy$$

Uranium238   Thorium 234

$$^{226}_{88}Ra \rightarrow {}^{222}_{86}Rn + {}^{4}_{2}\alpha + Energy$$

Radium 226   Radon 222

Beta Decay
A neutron within the nucleus of an isotope splits into a proton and a high energy electron.

- A new proton replaces the neutron
- The Mass number stays the same

Examples:

$$^{131}_{53}I \rightarrow {}^{131}_{54}Xe + {}^{0}_{-1}\beta + Energy$$

Iodine 131   Xenon 131

$$^{90}_{38}Sr \rightarrow {}^{90}_{39}Yt + {}^{0}_{-1}\beta + Energy$$

Strontium 90   Yttrium 90

Gamma Decay

When some isotopes decay through alpha or beta decay processes, the nucleus is left with a large amount of extra energy. This energy causes the nucleus to become unstable and move violently. It then releases this energy as a high energy electromagnetic wave in the gamma ray range. The isotope, however, does not change – the mass and atomic numbers stay the same.

Chapter 53. Half Life

Each isotope has a unique half-life. These can range from fractions of a second to billions of years.

- Half-life is the period of time it takes for the amount of a substance undergoing decay to decrease by half.

Since the activity (the number of clicks heard by a Geiger-Muller counter) is directly proportional to the number of isotopes in a substance that is decaying, it can also be expressed as:

- The half-life is the time taken for the activity to reduce by half

The table below shows how the percentage of the original isotope remains based on the number of half-lives that have passed. For carbon 14 the half-life is 5700 years. In a sample of 1000000 carbon-14 isotopes, on average, 500000 should be left after 5700 years, and then 250000 after a further 5700 years. After a further 5700 years that number will be 125000. It is important to remember that this number is only on average – in other words, with enough in the sample it should be this number, but since the decay is random, the number could be larger or smaller. For iGCSE you must remember that it is random, and to use the half-life in the way shown here.

Half Lifes	Years past	Fraction left	Percentage left	Number of C-14 isotopes left
1	0	1	100%	1000000
2	5730	0.5	50.0%	500000
3	11460	0.25	25.0%	250000
4	17190	0.125	12.5%	125000
5	22920	0.0625	6.3%	62500
6	28650	0.03125	3.1%	31250
7	34380	0.015625	1.6%	15625
8	40110	0.0078125	0.8%	7812

The graph below shows how the percentage of carbon 14 changes over time, as per the list above. The graph describes how the number halves and then halves again.

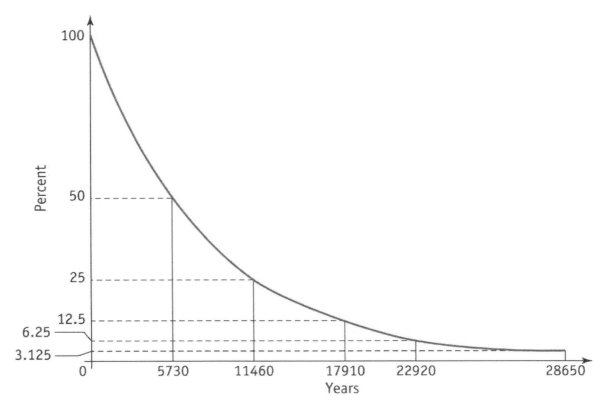

Figure 336 - Half-life graph for carbon 14

Following the example above, the following table show how the number of clicks, or the amount of the substance, changes over time:

Table 15 - Showing half-life in numbers for carbon 14

Number of half lives	Time elapsed (years)	Click rate (clicks per second)	Amount of substance left (number of atoms)
0	0	64000	128 million
1	5700	32000	64 million
2	11400	16000	32 million
3	17100	8000	16 million
4	22800	4000	8 million

Worked Example

In an experiment, a radioactive isotope's count rate (click rate), falls from 200 counts per second to 25 counts per second in 75 minutes, calculate it's half-life.

$$200\ counts \rightarrow (one\ half\ life) \rightarrow 100\ counts \rightarrow (2\ half\ lives) 50\ counts \rightarrow (3\ half\ lives) \rightarrow 25\ counts$$

So, there are 3 half lives that have passed until the count rate has changed from 200 to 25. As such a half-life must be 75/3, or **25 minutes**

It is also possible to calculate the count rate after a period of time. To do this 4 things are required:

1) The half-life

2) The initial count rate
3) The background count
4) The time duration

Worked Example

The half-life of a substance is 3 days. If the count rate starts at 2050 counts per minute (cpm), and the background radiation is 50cpm, what will the count rate be 9 days later?

First subtract the background rate from the 2050, leaving 2000 cpm from the substance itself.

$$2000 \; cpm \to (3 \; days) \; 1000 \; cpm \to (+ 3 \; days) \; 500 \; cpm \to (+3 \; days) \; 250 cpm$$

So, after 9 days, the count rate will be 250 counts per minute due to the substance itself, but we have to add back the background count of 50, so the answer will be 300 cpm.

53.2 Questions

1) A milk sample for Chernobyl containing Iodine-131, has an activity of 1600/litre. The activity of the sample was measured every 7 days. The results are shown in the table below:

Time (days)	0	7	14	21	28	35
Activity (units per litre)	1600	875	470	260	140	77

   a. Draw a graph of the activity against time
   b. Estimate the half-life of Iodine-131
2) Explain why some elements are radioactive and some are not
3) State the causes of background radiation
4) Americium 241 (Am) loses an alpha particle and decays into Neptunium. Produce and equation to show this (Am has an atomic number of 95)
5) Carbon 14 is radioactive and decays by beta emission. Its atomic number is 6.
   a. What new element is formed when carbon 14 decays
   b. Produce a symbol equation for this
6) Why would an isotope that emits alpha radiation be unsuitable as a tracer to monitor the heart
7) Radioactive iodine decays by beta emission, emitting gamma radiation. The following equation shows this:

$$^{131}_{53}I \to \; ^{A}_{Z}Xe + \; ^{0}_{-1}e$$

   a. What is the nucleon number and atomic number for this isotope of Xenon?
   b. The number of neutrons in the Xe nucleus
   c. The half-life of iodine 131 is 8 days. The total dose given to a patient emits 4×10^8 gamma rays per second. How many gamma rays will it emit after 24 days?

Chapter 54. Fission

Fission occurs when an isotope splits into smaller atoms. This can either occur spontaneously, because the isotope is unstable, or it can be induced by causing a stable isotope to become unstable. An example of nuclear fission is the decay into an alpha particle, as discussed above. A material that can easily be split, or go through fission, is called a fissile material.

Alpha decay is, however, just one of many ways that a heavy element can decay into lighter ones. A very heavy isotope, such as Uranium, can split into other, heavy, isotopes. These will then, in turn split into lighter ones, usually no lighter than Iron (Fe), which is the most radioactively stable of all elements.

In an induced fission reaction a neutron is "fired" at a very specific velocity (not too fast, not too slow) at a stable isotope, such as U-235, which then absorbs it and becomes U-236. This new isotope is very unstable and will split into two heavy "daughter" elements, and will additionally release 3 further neutrons.

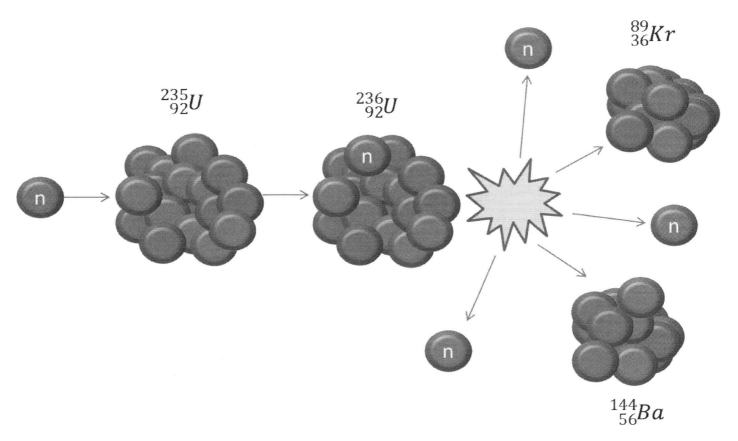

The released neutrons have a lot of energy, and thus move very rapidly. This means that they are unlikely to cause further reactions, unless they are slowed down, or unless there is a high density of other U-235 isotopes with which to react.

The nuclear reaction also releases a lot of energy that can easily transformed into heat.

The nuclear reaction for above can be written as:

$$^{235}_{92}U + ^{1}_{0}n \rightarrow ^{236}_{92}U \rightarrow ^{89}_{36}Kr + ^{144}_{56}Ba + 3^{1}_{0}n + energy + gamma\ radiation$$

Neutrons have an atomic number of 0 since there are no protons, and a mass number of 1 since it is a nucleon. Notice how the nucleon numbers add up to balance both sides of the equation. U-236 has 236 nucleons, and the Krypton, Barium and 3 neutrons all add up to 236 nucleons. The same is true for the atomic number, which adds up to 92.

If the neutrons go on to join other U-235 isotopes, these will cause the reaction to continue

54.1 Chain reactions

A chain reaction is a reaction where a reaction causes further reactions to occur, and these cause further reactions to occur, etc. In the case of a nuclear fission chain reaction, as outlined above, the released neutrons join further U-235 isotopes, which in turn become U-236 and then goes through fission, releasing further neutrons that then cause further reactions to take place.

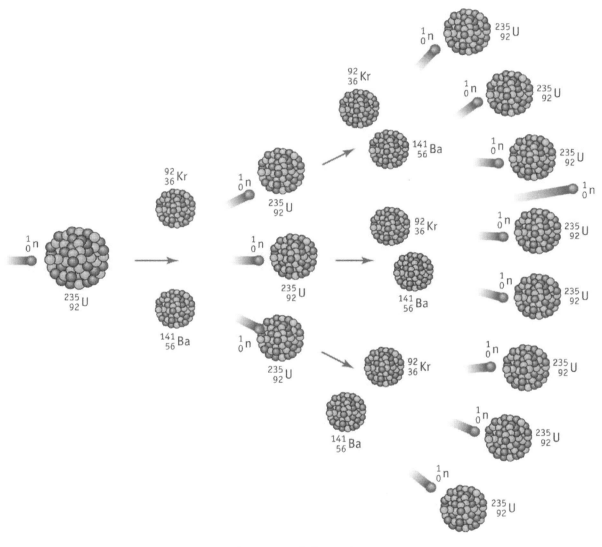

Figure 337 - Chain reaction

Uncontrolled nuclear chain reactions
If the above nuclear reaction is left uncontrolled, and if there is enough U-235, the result is a massive release of energy. This is known as an atomic bomb. These explosions are large enough to wipe out small cities and releases huge amounts of dangerous radiation.

Figure 338 - Atomic bombs are uncontrolled nuclear chain reactions

Controlled nuclear chain reactions (nuclear power stations)

Certain parameters must be met before a chain reaction can occur:

1) There is enough fissable material in a small enough space
2) The neutrons are travelling at the right speed
3) The neutrons can reach U-235 that has not yet been split
4) A chain reaction with the "right" number of neutrons reaching the next atom

By controlling these it is possible to control a nuclear reaction so that it is hot and can warm up water (or another fluid) that in turn can cause the spinning motion of a turbine and generate electricity.

A typical high-pressure water reactor is contained in a very strong lead and concrete reaction vessel so that no radiation is released into the outside world.

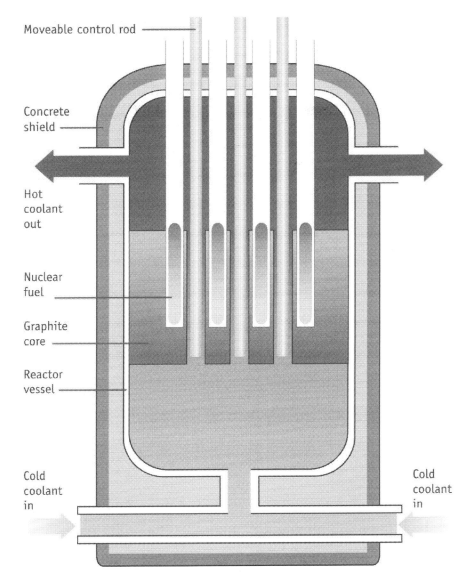

Figure 339 - Nuclear reaction vessel

Inside this reaction vessel there is set of technologies that control the above parameters.

Nuclear fuel rods
These rods are tubes filled with pellets that contain a carefully mixed combination of U-235 and other isotopes so that there is not risk of too much U-235 being in the same space. In a different design of powerstations the fuel used in Plutonium-239 (Pu-239); $^{239}_{94}Pu$.

These rods are about 4m long and stand upright in the reaction vessel. Surrounding these are moderator rods is a moderator.

Moderator
A moderator ensures that the neutrons from previous fission are slowed down to a velocity that is just right. Typical moderator materials are:

1) Graphite
2) Heavy water (Using ^2_1H to form water instead of "normal" ^1_1H hydrogen)
3) Lithium or Beryllium

Control Rods

Where the moderator reduces the kinetic energy of the neutron, the control rods completely absorb the neutron as it travels between the fuel rods. In normal use they ensure that only one of the 3 released neutrons go on to cuase further reactions. It is these rods that are raised and lowered to increase or decrease the rate of reaction by removing those neutrons that would cause the reaction to have a higher rate. Pushing the control rod further down into the reaction vessel, between the fuel rods, causes the reaction to slow down. If pushed in completely, the reaction is stopped.

Typical materials used to absorb neutrons are:

1) Boron
2) Silver-indium-cadmium alloys
3) Hafnium

Issues involved in nuclear power

Nuclear power has a bad name since it has the potential to be extremely dangerous. However, to date, very few people have died from direct damage from exposure to radiation from a nuclear power plant. The two largest nuclear disasters, Chernobyl and Fukushima, both were very dramatic and required huge areas to be evacuated, and they still are.

Figure 340 - Fukushima explosion on the left and the remnants of Chernobyl on the right

There are only presently 64 deaths that are directly related to the Chernobyl accident, most of those were emergency workers who worked to seal the nuclear core. There are no recorded deaths related to the Fukushima disaster, although it was a dramatic explosion in the containment vessel. There are, however, estimates of future deaths from thyroid cancer in both nuclear disasters, and from the fallout from Chernobyl, which reached across much of Europe.

The main issues with nuclear power involve how to store the spent fuel. This needs storing safely, away from terrorists and exposure to the elements for 10s of thousands of years. There are, in addition, lower level radioactive materials that from the nuclear power stations that need destroying, or storing, for longer periods, e.g. old pipes and overalls worn by the workers.

Nuclear power also requires a long time to come on-line with power. They are expensive to build and then need very careful maintenance and upkeep. At the end of their life, about 25-50 years, they need to be decommissioned gradually. This process is very costly too. Nuclear reactors use either Pu-239 (which is man-made, from Uranium), or U-235. U-235 exists in nature but is not renewable, and so will eventually run out.

54.2 Fusion

Nuclear fusion is the reverse of nuclear fission. In fusion, light elements travelling very quickly, such as Hydrogen and Helium, are joined together to make heavier ones, up to Iron, Fe (element 25, with 55 nucleons). To form heavier elements energy is supplied, not released. Nuclear fusion produces far more energy than nuclear fission for the same mass of material used, but, as yet, it is only possible for humans to produce uncontrolled nuclear fusion, as seen in hydrogen bombs, also known as thermonuclear bombs.

Figure 341 - Thermonuclear explosion

A thermonuclear explosion would wipe out a large city, such as London or New York.

A controlled fusion reaction requires the reaction to be continuous, using the released energy to stay hot. If the reaction stops for some reason, the temperature will drop and no further fusion will take place. As such, a fusion reaction would be safer than a fission reactor since when something went wrong in the reaction, it stops altogether.

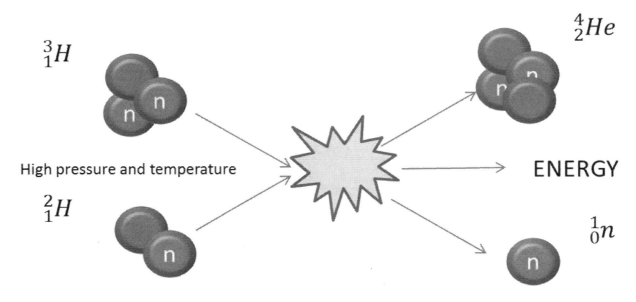

Figure 342 - Typical fusion reaction where neutron heavy hydrogen isotopes join together into helium

The sun and other stars are huge fusion reactors that have massive pressure inwards from the force of gravity that in turn causes heat and compression of the hydrogen until fusion starts to occur at very high temperatures. Attempts to simulate these circumstances so far on Earth have failed, but research is going on, on two concepts, one based on enclosing the high temperature hydrogen gas with enormous magnets around a toroid (a donut-like shape), and other based on high powered lasers. Once these technologies have become operational, the energy supply for the very long distant future is assured.

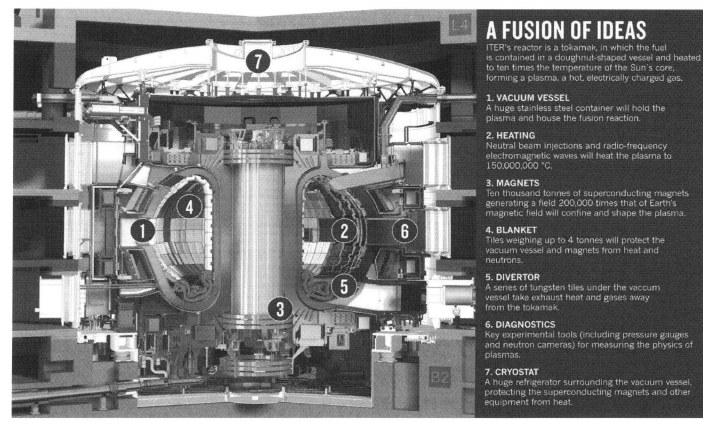

Figure 343 - Experimental Fusion reactor

54.3 Questions

1) U-235 is a fissile material
    a. what does this mean?
    b. If there is enough U-235 it may start a chain reaction
        i. In your owns words, what is a chain reaction?
        ii. Why does a chain reaction depend on how much material there is in one piece
2) List two advantages and two disadvantages of producing power using nuclear energy
3) Complete the following reaction:

$$^2_1H + ^3_1H \rightarrow ^?_?? + ^?_?n$$

4) How many protons and neutrons are there in $^{236}_{92}U$?
5) Copy and complete the sentences below, using the following words
   nucleus U-235 U-236 Plutonium-239
    a. Nuclear fission reactions occur when a _____ of _____ or _____ splits
    b. A nucleus of _____ in a nuclear reactor changes without fission into a nucleus of _____

6) Put the following statements in the right order to describe the chain reaction in a nuclear power station
    a. Two daughter isotopes are formed
    b. A U-236 atom splits
    c. A U-236 isotope is (briefly) formed

d. A neutron hits a U-235 nucleus
   e. Neutrons are released
   f. Energy is released
7) What slows down a fast neutrons, such as those released in a U-236 fission reaction in a nuclear power station, so that it can join U-235 atoms in another fuel rod?
8) What is used to completely stop a nuclear reaction in a nuclear power station?
9) What two fuels are used on nuclear power stations?

List of Figures

FIGURE 1 - A STANDARD 1KG MASS KEPT FROM CHANGING IN THE ATMOSPHERE BY BEING ENCASED IN A VACUUM 7
FIGURE 2 - STANDARD PREFIXES 7
FIGURE 3 - MASS BALANCE 9
FIGURE 4 - BEAM BALANCE 9
FIGURE 5 - DIGITAL STOPWATCH 11
FIGURE 6 - ANALOGUE STOPWATCH 11
FIGURE 7 - A PICTURE OF THE US STANDARD METRE BARS 11
FIGURE 8 - PARALLAX 13
FIGURE 9 - NEWTON METER 15
FIGURE 10 - FREE BODY DIAGRAM 15
FIGURE 11 - HOOKE'S LAW - EXTENSION IS PROPORTIONAL TO THE FORCE APPLIED 16
FIGURE 12 - GRAPH OF A STRING WHILE IT IS BEING STRETCHED WITH ITS ELASTIC LIMIT 17
FIGURE 13 - HOOKE'S LAW IS OBSERVED UNTIL THE ELASTIC LIMIT AND THEN THE LINE IS NO LONGER STRAIGHT. 17
FIGURE 14 - EXPERIMENTAL SETUP TO EXPLORE HOOKE'S LAW 19
FIGURE 15 - ISAAC NEWTON 1642-1727. DISCOVERER OF THE THEORY OF GRAVITY, INVENTOR OF ALGEBRA, THE TELESCOPE AND DEDUCTOR OF THE THREE LAWS OF MOTION. 21
FIGURE 16 - AREAS OF HIGH AND LOW G ON EARTH 22
FIGURE 17 - 3RD LAW ILLUSTRATION 29
FIGURE 18 - FORCE PAIRS IN SKATERS 30
FIGURE 19 - DISTANCE TIME GRAPH 33
FIGURE 20 - DISTANCE TIME GRAPH SHOWING CHANGES IN SPEED 34
FIGURE 21 – DISPLACEMENT=10KM AND DISTANCE=25KM 35
FIGURE 22 - DISPLACEMENT = 0 35
FIGURE 23 - VELOCITY - TIME GRAPH – THE AREAS UNDER THE GRAPH SHOWS THE DISTANCE TRAVELLED. USE STANDARD AREA CALCULATIONS TO WORK OUT. 36
FIGURE 24 - TERMS USED TO DESCRIBE A VELOCITY-TIME GRAPH 37
FIGURE 25 - STANDARD GOVERNMENT CALCULATIONS FOR STOPPING DISTANCES 38
FIGURE 26 - CENTRIPETAL FORCE ACTING ON A PLANET FROM THE GRAVITY OF A STAR. 39
FIGURE 27 - WHAT HAPPENS WHEN THE STRING SNAPS 39
FIGURE 28 - RELATING THE VELOCITY, PATH OF MOTION, CENTRIPETAL FORCE AND RADIUS OF THE CIRCULAR MOVEMENT 40
FIGURE 29 - AREA UNDER A GRAPH SHOWING THE CHANGE IN DISPLACEMENT BETWEEN TWO VELOCITIES. 42
FIGURE 30 - USING A MOMENT TO LOOSEN A NUT 49
FIGURE 31 - PRINCIPLE OF MOMENTS 51
FIGURE 32 - WORK OUT THE FORCE, F 51
FIGURE 33 - MULTIPLE MOMENTS IN EQUILIBRIUM 52
FIGURE 34 - WORK OUT THE DISTANCE, X 53
FIGURE 35 - WORK OUT THE MISSING DISTANCE, D 53
FIGURE 36 - CENTRE OF MASS OF AN IRREGULAR BODY IS WHERE ALL THE LINES OF ACTION MEET 54
FIGURE 37 - CENTRE OF MASS OF REGULAR OBJECTS 55
FIGURE 38 - STABILITY IS INCREASED BY WIDENING THE BASE 56
FIGURE 39 - BY ADDING MASS AT THE BOTTOM OF THE CYLINDER, THE CENTRE OF MASS IS LOWERED AND IT BECOMES MORE STABLE 56
FIGURE 40 - THE STATES OF EQUILIBRIUM 57
FIGURE 41 - A BUS AFFECTED BY TILTING DURING A TURN 58
FIGURE 42 - SEQUENCE OF BLOCKS WITH CENTRAL CENTRE OF GRAVITY 58
FIGURE 43 - SEQUENCE OF BLOCK WITH A LOW CENTRE OF GRAVITY 59
FIGURE 44 - FORMULA 1 CARS HAVE WIDE BASES AND A LOW CENTRE OF MASS 59
FIGURE 45 - TYPICAL PENDULUM 60
FIGURE 46 - WHAT MAKES THE FOSBURY FLOP SO EFFECTIVE? 61
FIGURE 47 - MEASURING THE DENSITY OF AN OBJECT USING THE AMOUNT OF WATER DISPLACED. 62

FIGURE 48 - BLAISE PASCAL	65
FIGURE 49 - PRESSURE AT THE BOTTOM IS HIGHER THAN AT THE TOP BECAUSE THE WEIGHT OF THE LIQUID ABOVE	66
FIGURE 50 - DAMS ARE BUILT STRONGER AT THE BOTTOM TO ENSURE THEY ARE STRONG ENOUGH TO WITHSTAND THE ADDITIONAL PRESSURE AT THE BOTTOM	67
FIGURE 51 - PRESSURE AT A DEPTH, H, IN A CYLINDER OF WATER	67
FIGURE 52 - A HYDRAULIC PUMP (IN REVERSE)	70
FIGURE 53 - HYDRAULIC PUMP	70
FIGURE 54 - THE DENSITY OF THE AIR REDUCES AS YOUR GO HIGHER. THE WEIGHT OF THAT AIR CAUSES THE ATMOSPHERIC PRESSURE AT LOWER LEVELS	71
FIGURE 55 - MERCURY BAROMETER	72
FIGURE 56 - AS A TUBE LEANS, THE HEIGHT IS DIMINISHED AND THE AMOUNT OF MERCURY IT SUPPORTS IS REDUCED	73
FIGURE 57 - MANOMETER	73
FIGURE 58 - EXAMPLE OF CONSERVATION OF ENERGY AND ENERGY TRANSFERS	76
FIGURE 59 - SNAKEY DIAGRAM FOR A FILAMENT LIGH BULB	77
FIGURE 60 - SANKEY DIAGRAM FOR AN ENERGY EFFICIENT LIGHT BULB	78
FIGURE 61 - ENERGY TRANSFERS IN A PENDULUM	79
FIGURE 62 - G.P.E. IS RELATIVE TO THE CHANGE IN HEIGHT.	80
FIGURE 63 - ENERGY TRANSFORMATION IN THE PRODUCTION OF ELECTRICITY	86
FIGURE 64 - NUCLEAR POWER PLANT. THE INITIAL HEAT TO THE STEAM GENERATOR IS CARRIED IN A CO_2 GAS SO THAT THE NUCLEAR FUEL DOES NOT CORRODE FROM WATER.	88
FIGURE 65 - TIDAL POWER DAM	89
FIGURE 66 - HYDROELECTRIC DAM	89
FIGURE 67 - THE HOOVER HYDROELECTRIC POWER STATION AND DAM IN THE USA	89
FIGURE 68 - SMALL WINDFARM IN RURAL SWEDEN	90
FIGURE 69 - SOME WAVE POWER DESIGNS USE THE UP AND DOWN MOTION TO PRODUCE ELECTRICITY	90
FIGURE 70 - SOLAR FARM	91
FIGURE 71 - GEOTHERMAL POWER PLANT IN ICELAND, WHERE THE EARTH'S CRUST IS THIN	93
FIGURE 72 - JAMES PRESCOTT JOULE	94
FIGURE 73 - CALCULATE THE POWER	95
FIGURE 74 - THE FRONT OF THE CAR CRUMPLES IN, REDUCING THE FORCE ON THE PASSENGERS	98
FIGURE 75 - EVOLUTION OF THE UNIVERSE FROM THE BEGINNING.	100
FIGURE 76 - A YOUNG STAR WITH ITS GAS DISC STILL AROUND IT	101
FIGURE 77 - ZOOMING OUT FROM EARTH TO THE SOLAR SYSTEM, OUR LOCAL STARS AND OUR PLACE IN THE MILKY WAY	103
FIGURE 78 - CONTINUING TO ZOOM OUT FROM THE MILKY WAY TO OUR LOCAL GROUP, THE SUPER CLUSTER WE ARE PART OF AND THEN FURTHER OUT AGAIN.	103
FIGURE 79 - FILAMENTS THAT FORM THE UNIVERSE. EACH DOT OF LIGHT IS A SUPER CLUSTER.	104
FIGURE 80 - STELLAR LIFE CYCLES FOR MOST STARS	104
FIGURE 81 - RED SUPERGIANT ELEMENT COMPOSITION JUST BEFORE COLLAPSE	105
FIGURE 82 - HOW THE DOPPLER EFFECT STRETCHES LIGHT WAVES	106
FIGURE 83 - EDWIN HUBBLE	106
FIGURE 84 - COSMIC BACKGROUND RADIATION IN THE MICROWAVE SPECTRUM	107
FIGURE 85 - CENTRIPETAL FORCE CAUSED BY GRAVITY CAUSING A SATELLITE TO ORBIT A PLANET	108
FIGURE 86 - JOHANNES KEPLER 1571-1630 IS CREDITED WITH THE PHYSICS OF SATELLITE MOTION	108
FIGURE 87 - THE SPACE AROUND EARTH IS FILLED WITH SATELLITES. THE RING IS WHERE ALL GEOSTATIONARY SATELLITES MUST ORBIT	109
FIGURE 88 - A COMET. THE TAIL ALWAYS POINTS AWAY FROM THE SUN	109
FIGURE 89 - THE LOCATIONS OF THE 8 PLANETS IN THE SOLAR SYSTEM - NOT TO SCALE	111
FIGURE 90 - COMETS HAVE ECCENTRIC ORBITS, COME IN FROM ALL KINDS OF ANGLES AND THEIR TAILS ALWAYS POINT AWAY FROM THE SUN	118
FIGURE 91 - BLACK BODY RADIATION	119
FIGURE 92 - HERTZSPRUNG-RUSSELL DIAGRAM	121
FIGURE 93 - BLACKBODY GRAPH	122
FIGURE 94 - HUBBLE'S LAW	124

Figure	Page
FIGURE 95 - THE THREE PHASES OF MATTER	126
FIGURE 96 - COMPARISONS BETWEEN THE THREE PHASES	127
FIGURE 97 - EXPLORING BROWNIAN MOTION. A GLASS CONTAINER IS FILLED WITH SMOKE. THE SMOKE PARTICLES CAN BE SEEN BUMPING OFF OTHER, INVISIBLE, AIR PARTICLES.	128
FIGURE 98 - A TYPICAL PATH OF A SMOKE PARTICLE WHEN OBSERVED IN THE APPARATUS SHOWN ABOVE	128
FIGURE 99 - THE THREE MAIN TEMPERATURE SCALES COMPARED	129
FIGURE 100 - A HEAT PUMP THAT REMOVES HEAT FROM THE COLD INSIDE OF A FRIDGE TO THE WARMER OUTSIDE	130
FIGURE 101 - HOW BOILING AND EVAPORATION DIFFER FROM EACH OTHER	131
FIGURE 102 - HOW A BIMETALLIC STRIP MIGHT BEND AND HOW THEY ARE SHOWN IN USE IN SCHOOLS	133
FIGURE 103 - FIRE ALARM USING A BIMETALLIC STRIP TO DETECT WHEN IT IS TOO HOT	134
FIGURE 104 - THERMOSTAT USING A BENT AND FLEXING BIMETALLIC STRIP TO CONTROL THE TEMPERATURE	134
FIGURE 105 - GAPS IN RAILS TO ALLOW EXPANSION	135
FIGURE 106 - ROLLER AT END OF BRIDGE TO ALLOW EXPANSION, SLOTS BETWEEN THE JOINS ALLOW THE PIECES TO SLIDE INTO AND OUT OF EACH OTHER	135
FIGURE 107 - HOOP IN PIPING TO ALLOW EXPANSION	136
FIGURE 108 - LOOSE CABLES ON PYLONS ALLOW FOR EXPANSION AND CONTRACTION DUE TO HEAT	136
FIGURE 109 - GRAPH SHOWING HOW THE DENSITY OF WATER IS HIGHEST AT 4C	137
FIGURE 110 - IN GLASS MERCURY (OR ALCOHOL) THERMOMETER	140
FIGURE 111 - TYPICAL THERMOCOUPLE	141
FIGURE 112 - ROBERT BOYLE 27 JANUARY 1627 - 31 DECEMBER 1691	142
FIGURE 113 - JOSEPH LOUIS GAY-LUSSAC, 6 DECEMBER 1778 – 9 MAY 1850	144
FIGURE 114 - MEASURING SPECIFIC HEAT CAPACITY OF WATER	148
FIGURE 115 - HOW HEAT IS TRANSFERRED BETWEEN THE STATES	150
FIGURE 116 - A COOLING CURVE - THE TEMPERATURE DROPS UNTIL IT CHANGES PHASES AND THEN LEVELS OFF UNTIL IT HAS COMPLETELY SOLIDIFIED. THE TEMPERATURE THEN CONTINUES TO DROP.	151
FIGURE 117 - PHASE CHANGE DIAGRAM, SHOWING THE STAGES WHERE LATENT HEAT IS USED TO CHANGE THE PHASES	152
FIGURE 118 - EXPERIMENT TO FIND THE LATENT HEAT OF MELTING OF ICE TO WATER	153
FIGURE 119 - MEASURING SPECIFIC LATENT HEAT OF BOILING	154
FIGURE 120 - HEAT TRANSFER IN ELECTRICALLY CONDUCTING METALS VIA ELECTRON MOVEMENT	156
FIGURE 121 - HOW A "SEA" OF ELECTRONS MOVES HEAT AWAY FROM THE HOT AREA	157
FIGURE 122 - PRACTICAL TO ILLUSTRATE DIFFERENT RATES OF CONDUCTION	158
FIGURE 123 - WINDS BLOW ONTO LAND DURING THE DAY, AND OUT TO SEA AT NIGHT	159
FIGURE 124 - CONVECTION CURRENTS CARRY THE HEAT UP, ACROSS AND DOWN A ROOM FROM A HEATING ELEMENT	160
FIGURE 125 - COLD, MORE DENSE, AIR FROM THE FREEZER FALLS, DISPLACING THE WARMER AIR UP TO THE FREEZER COMPARTMENT	160
FIGURE 126 - HOT WATER HEATING USING CONVECTION	161
FIGURE 127 - HEAT RADIATING FROM A CUP VIA INFRARED RADIATION	161
FIGURE 128 - MATT BLACK IS BOTH THE BEST RADIATOR AND ABSORBER OF RADIATED HEAT	162
FIGURE 129 - HOUSES IN HOT COUNTRIES ARE PAINTED WHITE TO REFLECT THE HEAT	162
FIGURE 130 - VACUUM FLASKS USE MANY TECHNIQUES TO KEEP THE HEAT IN	163
FIGURE 131 - THERMOGRAPHY USED IN MEDICINE AND IN CHECKING HOUSE INSULATION	164
FIGURE 132 - GREENHOUSES ALLOW INFRARED LIGHT TO ENTER, BUT NOT TO ESCAPE	165
FIGURE 133 - WAVE TERMINOLOGY	167
FIGURE 134 - WAVELENGTHS MEASURED AS VARIOUS PARTS OF THE WAVE	168
FIGURE 135 - MORE TYPICAL PLACES TO MEASURE WAVELENGTHS	168
FIGURE 136 - COMPRESSIONS IN A GAS WITH THE INDIVIDUAL PARTICLES ONLY MOVE FROM LEFT TO RIGHT. THE WAVE TRAVELS TO THE RIGHT.	169
FIGURE 137 - RAREFACTIONS AND COMPRESSIONS IN A SOUND WAVE	170
FIGURE 138 - RAREFACTION AND COMPRESSION IN A SLINKY	170
FIGURE 139 - SLINKY BEING SHAKEN FROM SIDE TO SIDE TO GENERATE A TRANSVERSE WAVE	170
FIGURE 140 - WATER WAVES, WHERE THE VARYING HEIGHT OF THE WAVE SHOWS THE UP AND DOWN MOTION OF THE OSCILLATIONS	171
FIGURE 141 - NORMAL TO A STRAIGHT LINE SEGMENT	171

Figure	Description	Page
FIGURE 142	NORMAL TO A CURVE.	172
FIGURE 143	RIPPLE TANK	173
FIGURE 144	AN INCIDENT RAY, REFLECTING OFF A SURFACE.	174
FIGURE 145	THE TWO ANGLES ARE CALLED THE "INCIDENT ANGLE" AND THE "REFLECTED ANGLE" THEY ARE OFTEN REPRESENTED BY THE GREEK LETTER, THETA.	174
FIGURE 146	CONCAVE AND CONVEX MIRRORS REFLECTING RAYS	174
FIGURE 147	WAVE FRONT INTERACTION WITH BARRIERS, SUCH AS MIRRORS	175
FIGURE 148	SPEED CHANGE AS A WAVE MOVES THROUGH DIFFERENT DEPTHS OF WATER	176
FIGURE 149	NOTE HOW THE WAVES CHANGE BOTH DIRECTION AND WAVELENGTH AS THE APPROACH LAND	176
FIGURE 150	REFRACTION IN WATER, SHOWING CHANGE OF DIRECTION AND WAVELENGTH	177
FIGURE 151	DIFFRACTION AT AN EDGE, OR WHEN PASSING THROUGH A WIDE OPENING	177
FIGURE 152	DIFFRACTION THOUGH AND OPENING WHERE THE WIDTH IS ABOUT THE SIZE OF THE WAVELENGTH	178
FIGURE 153	DIFFRACTION AT AN INLET	178
FIGURE 154	THE TWO TYPES OF REFLECTION OF LIGHT	181
FIGURE 155	FIRST LAW OF REFLECTION - ANGLE OF INCIDENCE = ANGLE OF REFLECTION	182
FIGURE 156	REFLECTION IN A MOUNTAIN LAKE	183
FIGURE 157	THE LOCATION OF A VIRTUAL IMAGE IN A PLANE MIRROR	183
FIGURE 158	IMAGE OF A VASE	184
FIGURE 159	DRAWING THE LINES TO LOCATE THE VIRTUAL IMAGE	184
FIGURE 160	THE DISTANCES OF THE IMAGE AND REAL OBJECT TO THE MIRROR ARE THE SAME	185
FIGURE 161	REFRACTION FROM FAST TO SLOW MEDIA	186
FIGURE 162	REFRACTION THROUGH A BLOCK OF GLASS	187
FIGURE 163	REFRACTION IN A SOLID BLOCK OF GLASS	188
FIGURE 164	EXPLORING THE ANGLES IN REFRACTION	189
FIGURE 165	CRITICAL ANGLE IN A SEMI-CIRCULAR BLOCK	193
FIGURE 166	TOTAL INTERNAL REFLECTION	194
FIGURE 167	- TIR IN THE SURFACE OF WATER - NOTE THE REFLECTION OF THE TORTOISE IN THE SURFACE	195
FIGURE 168	USING TWO PRISMS THAT HAVE 45 DEGREE ANGLES TO THE VIEWER ALLOWS SOMEONE TO VIEW AN IMAGE HIGHER UP THAT IS ALSO THE RIGHT WAY UP	196
FIGURE 169	HOW CAT'S EYES WORK USING TIR	197
FIGURE 170	TOTAL INTERNAL REFLECTION IN AN OPTICAL FIBRE	197
FIGURE 172	OPTICAL FIBRES	198
FIGURE 172	TOTAL INTERNAL REFLECTION IN A PERSPEX BLOCK	198
FIGURE 173	CONVEX LENSES	199
FIGURE 174	CONCAVE, OR DIVERGING, LENS	199
FIGURE 175	REDUCED IMAGE OF LARGER OBJECT IN A CONCAVE MIRROR	200
FIGURE 176	CONVERGING LENS WHERE THE OBJECT IS BEYOND 2F	200
FIGURE 177	COVERGING LENS THERE THE OBJECT IS BETWEEN F AND 2F	201
FIGURE 178	CONVERGING LENS WHERE THE OBJECT IS CLOSER THAN THE FOCAL POINT, F	201
FIGURE 179	DISPERSION (THROUGH REFRACTION) IN A PRISM	204
FIGURE 180	DIAGRAM OF THE HUMAN EYE	205
FIGURE 181	IMAGES AT THE BACK OF THE EYE ARE LATERALLY INVERTED	206
FIGURE 182	MYOPIC EYEBALL	207
FIGURE 183	HYPEROPIC EYEBALL	207
FIGURE 184	LENSES USED TO CORRECT SHORT- AND LONG SIGHTEDNESS	208
FIGURE 185	THE WHOLE ELECTROMAGNETIC SPECTRUM	210
FIGURE 186	RADIOTHERAPY USES GAMMA RAY TO TREAT CANCER	211
FIGURE 187	SOUNDS WAVES - SHOWING RAREFACTION AND COMPRESSIONS.	213
FIGURE 188	SHOCK WAVES, SUCH AS THESE FROM THE GUNS OF A BATTLE SHIP, ARE SINGLE, VERY HIGHLY COMPRESSED AREAS OF AIR PARTICLES THAT CAN BE DAMAGING TO STRUCTURES AND PEOPLE.	213
FIGURE 189 - HIGH PITCH, QUIET	FIGURE 190 - HIGH PITCH, LOUD	215
FIGURE 191 - LOW PITCH, QUIET	FIGURE 192 - LOW PITCH, LOUD	216
FIGURE 193	THE MODEL OF THE ATOM	217

Figure	Description	Page
FIGURE 194	HOW TO VISUALISE MAGNETIC DOMAINS AND THEIR MAGNETIC ORIENTATION	219
FIGURE 195	HOW PERMANENT MAGNETS AFFECT EACH OTHER	220
FIGURE 196	STROKING A MAGNETIC MATERIAL WITH A PERMANENT MAGNET WILL CAUSE IT TO BECOME A MAGNET	221
FIGURE 197	A COIL WILL PRODUCE A MAGNETIC FIELD THAT WILL MAKE ANY MAGNETIC MATERIAL INSIDE A MAGNET	222
FIGURE 198	TRACING THE MAGNETIC FLUX LINES USING IRON FILINGS AND PLOTTING COMPASSES	223
FIGURE 199	MAGNETS THAT ATTRACT. NOTE HOW THE FIELD LINES JOIN TOGETHER	224
FIGURE 200	REPELLING MAGNETS. NOTE HOW THE FIELD LINES AVOID EACH OTHER	224
FIGURE 201	THE EARTH'S GEOGRAPHIC NORTH POLE IS A MAGNETIC SOUTH POLE	225
FIGURE 202	PLOTTING THE MAGNETIC FIELD AROUND A MAGNET	226
FIGURE 203	NOTE THE CONCENTRIC FLUX LINES	226
FIGURE 204	RIGHT HAND THUMB RULE	227
FIGURE 205	MAGNETIC FLUX LINES IN A COIL	228
FIGURE 206	THE RIGHT HAND COIL CLASP RULE - SHOWS THE DIRECTION OF AN ELECTROMAGNETIC FIELD PRODUCED BY A COIL OF WIRE	229
FIGURE 207	ELECTROMAGNET USED IN A RELAY	230
FIGURE 208	DOOR BELL USING AN ELECTROMAGNET	231
FIGURE 209	CIRCUIT BREAKER USING AN ELECTROMAGNET	232
FIGURE 210	PERSPEX AND CLOTH	233
FIGURE 211	POLYTHENE ROD WITH A CLOTH	233
FIGURE 212	OPPOSITE CHARGES ATTRACT EACH OTHER	234
FIGURE 213	SAME CHARGES REPEL EACH OTHER	234
FIGURE 214	ELECTRIC FIELD LINES	235
FIGURE 215	ATTRACTION AND REPULSION BETWEEN POINT CHARGES.	235
FIGURE 216	ELECTRIC FIELD LINES BETWEEN PARALLEL PLATES	236
FIGURE 217	CHARGE IS INDUCED ON THE SURFACE OF THE BALLOON BY THE ATTRACTION OF THE ELECTRONS TO THE ROD FROM THE BALLOON	236
FIGURE 218	LIGHTNING IS CAUSED BY THE EARTHING OF CHARGE FROM CLOUDS TO THE EARTH	237
FIGURE 219	FRICTION IN PIPES CAN CAUSE SPARKS.	237
FIGURE 220	VAN DER GRAAF GENERATORS CAUSE CHARGED PARTICLES TO BUILD UP ON THE SURFACE OF THE LARGE GLOBE.	238
FIGURE 221	HAIR STICKS ON END AS THE ELECTRONS FLOW TO THE END OF THE STRANDS OF THE HAIR AND REPEL EACH OTHER	239
FIGURE 222	CHARLES-AUGUSTIN DE COULOMB	240
FIGURE 223	A POSITIVELY CHARGED CAR ATTRACTS NEGATIVELY CHARGED PAINT PARTICLES	241
FIGURE 224	SMOKE PARTICLES ARE NEGATIVELY CHARGED AND THEN ATTRACTED TO POSITIVELY CHARGED COLLECTING PLATES	242
FIGURE 225	NEGATIVELY CHARGED INK DROPLETS ARE DIRECTED BY DEFLECTION PLATES TO THE RIGHT LOCATION ON PAPER	242
FIGURE 226	HOW STATIC CHARGE ATTRACTS TONER TO THE RIGHT LOCATION TO FORM AN IMAGE IN A PHOTOCOPIER	243
FIGURE 227	FLOW OF ELECTRONS	245
FIGURE 228	CONVENTIONAL CURRENT - FROM POSITIVE TO NEGATIVE	245
FIGURE 229	IN THE SERIES CIRCUIT ABOVE, THE ELECTRONS SHARE THEIR ENERGY BETWEEN THE BULBS, BUT THE CURRENT DOES NOT CHANGE	249
FIGURE 230	IN THE PARALLEL CIRCUIT ABOVE, THE ELECTRONS CAN CHOOSE TWO ROUTES BACK TO THE CELL.	250
FIGURE 231	GEORG OHM - USING EQUIPMENT OF HIS OWN CREATION, OHM FOUND THAT THERE IS A DIRECT PROPORTIONALITY BETWEEN THE POTENTIAL DIFFERENCE (VOLTAGE) APPLIED ACROSS A CONDUCTOR AND THE RESULTANT ELECTRIC CURRENT	255
FIGURE 232	CIRCUIT TO EXPLORE OHM'S LAW	257
FIGURE 233	SIMPLE "OHMIC" RESPONSE OF A RESISTOR TO A POTENTIAL DIFFERENCE ACROSS IT	258
FIGURE 234	HOW THE VOLTAGE CHANGES THE CURRENT IN A DIODE	259
FIGURE 235	THE CHANGE IN GRADIENT OF THE VOLTAGE VS CURRENT GRAPH SHOWS HOW, WHEN A BULB IS HOT, THE RESISTANCE INCREASES.	259
FIGURE 236	THIS IS THE MORE COMMON WAY OF LOOKING AT HOW A DIODE WORKS. WHEN THE VOLTAGE REACHES +0.7V, THE DIODE LETS THE CURRENT THROUGH	260

FIGURE 237 - THIS DIAGRAM SHOWS HOW, WHEN THE LIGHT BULB IS HOT, THE RESISTANCE INCREASES DRAMATICALLY, MAKING IT DIFFICULT FOR THE CURRENT TO FLOW. 260
FIGURE 238 - JAMES WATT, THE SCOTTISH INVENTOR AFTER WHOM THE UNIT FOR POWER IS NAMED. ONE OF THE INSTIGATORS OF THE INDUSTRIAL REVOLUTION 270
FIGURE 239 - TYPICAL FUSES 274
FIGURE 240 - A UK THREE-PINI MAINS SOCKET PLUG 275
FIGURE 241 - LOCATIONS OF THE WIRES AND PARTS OF A PLUG 275
FIGURE 242 - SAFELY WIRED 275
FIGURE 243 – A DOUBLE INSULATED APPLIANCE HAS THE APPLIANCE MADE OF AN INSULATED MATERIAL AND THE CABLES ARE NEVER EXPOSED 277
FIGURE 244 - BEFORE TOUCHING A VICTIM, TURN OFF THE POWER 279
FIGURE 245 - MOVING A WIRE BETWEEN TWO MAGNETS WILL INDUCE A CURRENT IN THE WIRE 280
FIGURE 246 - JOHN AMBROSE FLEMING 1849-1945 281
FIGURE 247 - FLEMING'S RIGHT-HAND RULE, OR THE GENERATOR RULE 282
FIGURE 248 - STEP 1, THE MAGNET IT PUSHED INTO THE COIL 283
FIGURE 249 - STEP 2, THE MAGNET IS LEFT STATIONARY INSIDE THE COIL. NO CURRENT IS INDUCED 283
FIGURE 250 - STEP 3, THE MAGNET IS PULLED OUT. CURRENT IS INDUCED IN THE OPPOSITE DIRECTION TO THAT IN STEP 1 283
FIGURE 251 - LAST STEP, THE MAGNET IS OUTSIDE THE COIL, AT REST, AND NO CURRENT IS THUS INDUCED. 284
FIGURE 252 - MICHAEL FARADAY (1791-1867) 285
FIGURE 253 - WHEN THE MAGNET ENTERS THE COIL, THE CURRENT WILL PRODUCE A SOUTH POLE TO REPEL THE MAGNET, WHEREAS, WHEN THE MAGNET IS PULLED OUT, THE INDUCED CURRENT WILL PRODUCE A MAGNET THAT ATTRACTS, OR OPPOSES, THE MAGNET BEING REMOVED. 286
FIGURE 254 - FORCE EXPERIENCED BY A WIRE WITH ITS CURRENT GOING INTO THE PAPER. THE FLUX LINES BUILD UP ON THE SIDE WHERE THE DIRECTIONS OF THE TWO MAGNETIC FIELDS, PUSHING THE WIRE DOWNWARDS. 290
FIGURE 255 - FLEMING'S LEFT-HAND RULE, OR THE MOTOR EFFECT RULE 290
FIGURE 256 - TURNING FORCE ON A COIL IN A MAGNETIC FIELD 291
FIGURE 257 - THE GRAPH SHOWS HOW THE CURRENT CHANGES DIRECTION REGULARLY OVER TIME. THIS IS CALLED AN ALTERNATING CURRENT 293
FIGURE 258 - AN GENERATOR 294
FIGURE 259 - THE MOTOR EFFECT. THE COMMUTATOR ROTATES WITH THE COIL AS PART OF THE SAME ASSEMBLY. AS IT ROTATES, THE BRUSHES ALTERNATELY TOUCH THE OTHER COMMUTATOR, CAUSING THE CURRENT TO CHANGE DIRECTION IN THE COIL, THUS CHANGING THE DIRECTION OF THE FORCE IN IT 295
FIGURE 260 - HOW THE COMMUTATOR CHANGES THE DIRECTION OF THE FLOW OF CURRENT 296
FIGURE 261 - A VARYING, OR CHANGING, MAGNETIC FIELD WILL INDUCE A CURRENT IN A NEIGHBOURING WIRE 298
FIGURE 262 - A STEP DOWN TRANSFORMER - NOTICE HOW THE PRIMARY COIL HAS MORE WINDS THAN THE SECONDARY COIL. 299
FIGURE 263 - STEP DOWN TRANSFORMER 300
FIGURE 264 - USING TRANSFORMERS IN ENERGY TRANSMISSION TO REDUCE LOSS THROUGH HEATING IN THE WIRES 304
FIGURE 265 - LARGE SUBSTATION 305
FIGURE 266 - LOCAL SUBSTATION 305
FIGURE 267 - CONTROVERSIAL LOCATION FOR A PYLON RUN 305
FIGURE 268 - CIRCUIT SYMBOL FOR A DIODE 309
FIGURE 269 - DIODES 310
FIGURE 270 - IN THIS CIRCUIT, THE CURRENT IS ALLOWED THROUGH THE DIODE AND THE BULB WILL LIGHT UP 310
FIGURE 271 - THE DIODE RESISTS THE CURRENT AS IT IS PLUGGED OPPOSING THE CONVENTIONAL CURRENT 310
FIGURE 272 - IN THIS CIRCUIT, ONLY THE CURRENT THAT FLOWS IN THE DIRECTION OF THE DIODE IS LET THROUGH. THE POSITIVE PEAKS ARE THE ONES LEFT 311
FIGURE 273 - IN THIS CONFIGURATION, THE OUTPUT SIGNAL IS OUTPUT AS A FULLY RECTIFIED DC LINK, BUT IT STILL IS NOT SMOOTH 311
FIGURE 274 - A SMOOTHED, RECTIFIED OUTPUT 312
FIGURE 275 - FULL-WAVE RECTIFICATION 312
FIGURE 276 - SMOOTHED OUTPUT USING A CAPACITOR 312
FIGURE 277 – NPN TRANSISTOR CIRCUIT SYMBOL 312
FIGURE 278 – AN NPN TRANSISTOR 313

FIGURE 279 - IN THIS CIRCUIT THE 19.9MA IS ADDED TO THE 0.1MA TO FORM A 20MA CURRENT	314
FIGURE 280 - SOME TRANSISTORS	314
FIGURE 281 - WHEN A CURRENT FLOWS BETWEEN THE PROBES, A CURRENT WILL FLOW THROUGH THE COLLECTOR TOO, LIGHTING THE BULB	314
FIGURE 282 - LDR CIRCUIT SYMBOL	315
FIGURE 283 - SIMPLE POTENTIAL DIVIDER THAT CAN BE USED TO TURN ON A STREET LAMP WHEN IT GOES DARK	316
FIGURE 284 - TIME DELAY CIRCUIT	317
FIGURE 285 - THERMISTOR CIRCUIT SYMBOL	317
FIGURE 286 - AN ALARM THAT SETS OF A BUZZER WHEN THE TEMPERATURE IS TOO HIGH	318
FIGURE 287 - LED CIRCUIT SYMBOL	318
FIGURE 288 - THIS CIRCUIT TURNS ON AN LED WHEN THE TEMPERATURE RISES TO A CERTAIN LEVEL.	319
FIGURE 289 - EXAMPLE OF AN ANALOGUE SIGNAL	321
FIGURE 290 - EXAMPLE OF A DIGITAL SIGNAL	321
FIGURE 291 - USING SWITCHES FOR AN AND CIRCUIT	323
FIGURE 292 - USING TRANSISTORS TO PROVIDE AN AND CIRCUIT	324
FIGURE 293 - AND SYMBOL	324
FIGURE 294 - A TRANSISTOR VERSION OF AN OR GATE	325
FIGURE 295 - OR SYMBOL	325
FIGURE 296 - NOT SYMBOL	325
FIGURE 297 - THE LONG VERSION OF A NAND GATE	326
FIGURE 298 - LOGICAL NAND GATE - NOTICE THE CIRCLE	326
FIGURE 299 - LONG VERSION OF A NOR GATE	327
FIGURE 300 - LOGICAL NOR GATE	327
FIGURE 301 - CRO WITH A TRACE ON ITS SCREEN	331
FIGURE 302 - THERMIONIC EMISSION - ELECTRONS "BOIL OFF" A FILAMENT	331
FIGURE 303 - THE ANODE ATTRACT THE THERMIONICALLY EMITTED ELECTRONS, CAUSING A STREAM OF THEM THROUGH A HOLE TOWARDS THE RIGHT	332
FIGURE 304 - THE RAY IS DEFLECTED UP OR DOWN DEPENDING ON THE INPUT VOLTAGE	332
FIGURE 305 - DEFLECTOR PLATES ATTRACT AND REPEL CATHODE RAY	333
FIGURE 306 - THE COMPLETE ASSEMBLY AS USED IN AN OCR. THE X-PLATES CAUSE A DEFLECTION IN THE HORIZONTAL PLANE AND THE Y-PLATES IN THE VERTICAL.	333
FIGURE 307 - A TYPICAL OSCILLOSCOPE	334
FIGURE 308 - ERNEST RUTHERFORD	338
FIGURE 309 - PLUM PUDDING MODEL WITH ELECTRONS BAKED INTO A POSITIVELY CHARGED ATOM	338
FIGURE 310 - GEIGER AND MARSDEN'S APPARATUS. ALPHA PARTICLES (FROM R) SCATTER OFF THE FOIL (F). THE MICROSCOPE (M) IS ROTATED AROUND THE CYLINDRICAL BOX (B) TO COUNT SCATTERING AT ANY ANGLE	339
FIGURE 311 - DEFLECTIONS OF ALPHA PARTICLES THAT HIT A "PLUM PUDDING" ATOM AT THE TOP, AND WHAT THEY FOUND, AT THE BOTTOM	340
FIGURE 312 - THE PATHS THAT THE ALPHA PARTICLES WERE OBSERVED TAKING	341
FIGURE 313- ATOMIC STRUCTURE, AS SEEN IN A CARBON ATOM	342
FIGURE 314 - HOW TO WRITE DOWN A NUCLIDE	343
FIGURE 315 - BUILDING AN ATOM - HYDROGEN WITHOUT AN ELECTRON, AS AN ION	344
FIGURE 316 - ADDING AN ELECTRON BALANCES THE PROTON CHARGE WITH THE ELECTRONS' OPPOSITE CHARGE	344
FIGURE 317 - ADDING A NEUTRON TO THE NUCLEUS DOESN'T ADD CHARGE, BUT DOES CREATE A AN ISOTOPE OF HYDROGEN	345
FIGURE 318 - ADDING ANOTHER NEUTRON STILL KEEPS IT AS HYDROGEN, BUT WITH TWO NEUTRONS. ANOTHER ISOTOPE OF HYDROGEN	345
FIGURE 319 - ADDING ANOTHER PROTON CHANGES THE HYDROGEN ELEMENT TO HELIUM. ADDING AN ELECTRON ENSURES IT IS NOT CHARGED	345
FIGURE 320 - ALPHA PARTICLE DECAY FROM A LARGER ATOM	348
FIGURE 321 - ALPHA PARTICLE, TWO PROTONS (+) AND TWO NEUTRONS (N)	348
FIGURE 322 - BETA PARTICLE EMISSION FROM AN ATOM	349
FIGURE 323 - GAMMA RAY EMISSION FROM THE NUCLEUS OF AN ATOM	349
FIGURE 324 - PENETRATIVE POWER OF DIFFERENT RADIATION TYPES	350

FIGURE 325 - HOW A MAGNETIC FIELD AND CHARGED PLATES AFFECT THE MOVEMENT OF A RADIOACTIVE PARTICLE 351
FIGURE 326 - DEFLECTION PATHS OF DIFFERENT PARTICLES PASSING THROUGH A MAGNETIC FIELD 352
FIGURE 327 - GOLD LEAF ELECTROSCOPE 354
FIGURE 328 - GM TUBE 355
FIGURE 329 - CLOUD CHAMBER 355
FIGURE 330 - SPARK COUNTER... 356
FIGURE 331 - RADIO THERAPY 357
FIGURE 332 - MEASURING THE THICKNESS OF PAPER 357
FIGURE 333 - IRRADIATED OPERATING EQUIPMENT 358
FIGURE 334 - RADIATION HAZARD SYMBOL.. 359
FIGURE 335 - WEARING RADIATION SUITS.. 359
FIGURE 336 - HALF-LIFE GRAPH FOR CARBON 14 363
FIGURE 337 - CHAIN REACTION 366
FIGURE 338 - ATOMIC BOMBS ARE UNCONTROLLED NUCLEAR CHAIN REACTIONS 367
FIGURE 339 - NUCLEAR REACTION VESSEL.. 368
FIGURE 340 - FUKUSHIMA EXPLOSION ON THE LEFT AND THE REMNANTS OF CHERNOBYL ON THE RIGHT 369
FIGURE 341 - THERMONUCLEAR EXPLOSION.. 370
FIGURE 342 - TYPICAL FUSION REACTION WHERE NEUTRON HEAVY HYDROGEN ISOTOPES JOIN TOGETHER INTO HELIUM 371
FIGURE 343 - EXPERIMENTAL FUSION REACTOR 372

Index

absorbers, 162
AC generators, 293
acceleration, 15
Acceleration, 32
Activity, 347
Alpha Decay, 360
Alpha Particle, 348
alpha particles, 338
alternating current, 293
alternator, 293
Ampere, 7
analogue, 321
AND, 323
Atmospheric Pressure, 72
Atomic number, 343
attenuation, 321
Balanced forces, 26
bell jar, 214
Beta Decay, 360
Beta particles, 348
bimetallic strip, 134, 137
Bi-metallic strips, 133
binary, 321
Biofuels, 92
BIOFUELS, 91
Boiling, 129, 131
Boyle's Law, 142
Brownian Motion, 127
Buckling, 135
Calibration, 140
Capacitors, 309
Car Painting, 241
Carbon 14, 343
Carbon 14 Dating, 357
Cathode ray oscilloscopes, 331
cats eyes, 197
Centre of mass, 54
centrifugal, 40
centripetal, 39
charge induction, 236
Chemical, 75
Circuit breakers, 231
Circuit Breakers, 274
Circular Motion, 39
COAL, 87

Collisions, 97
Compression, 15
compressions, 169
Concave Lenses, 199
Conduction, 156
Conductors, 157, 234
Conservation of Energy, 75
Conservation of Momentum, 98
continuous, 321
Convection, 158
conventional current, 245
Convex Lenses, 199
Coulomb, 241
Crests, 167
critical angle, 194
CRO, 331
Current, 244
current gain, 313
dark ages, 100
DC motor, 294
decay, 343, 347, 351, 357, 360
Deflection, 350
Density, 62
Diffraction, 177
Diffuse reflection, 181
Digital, 321
Diminished, 200
Diodes, 309
dispersion, 204
displacement, 167
Displacement, 34
Distance-time graphs, 33
earthed, 237
Earthing, 277
Efficiency, 78
Electric bells, 230
Electrical, 75
electrical accidents, 278
electrical interference, 321
Electrical power, 270, 271
Electricity generation, 86
Electromagnet, 228
electromagnetic induction, 280, 292
electromagnetic spectrum, 210

Electromotive Force, 246
electron, 338
Electrons, 218
Electrostatic Dust Precipitators, 241
EMF, 246, 294
Emitters, 162
Energy, 75
Energy transfer diagrams, 77
Energy transferred, 268
equilibrium, 54
equilibrium position, 167
Evaporation, 129, 130
expand, 133
Expansion of liquids, 136
Faraday's Law, 285
Fire Alarm, 134
Fleming's left had rule, 351
fluid, 156
flux lines, 220
force constant, 19
Forces, 15
Frequency, 167
Friction, 15
fulcrum, 49
Fuses, 274
galaxy, 102
Gamma Decay, 361
Gamma ray, 349
GAS, 87
generator, 280
Geothermal, 92
GEOTHERMAL, 91
Gravitation field strength, 22
gravitational potential energy, 80
gravity, 39
Half Life, 362
hard magnetic material, 220
Heat, 75
Heat Transfer, 156
Hooke, 16
Hydroelectric, 92
Hydroelectric power, 89
Impulse, 97
Incandescence, 181

induced charge, 240
Induced Current, 282
Inkjet Printers, 242
input, 309
insulators, 233
Insulators, 157
Internal Energy, 128
Inverted, 201
Ionisation, 353
irradiation, 358
isotope, 342, 343
Kelvin, 7, 129, 139, 141, 146
Kilogram, 7
Kinetic, 75
kinetic energy, 81
kinetic molecular model, 126, 127
Latent Heat, 150
LDR, 315
LED, 318
Left Hand rule, 290
Lenz's Law, 285
Lift, 15
Light, 75
light years, 100
Light-dependent resistor, 315
light-emitting diode, 318
line of action, 57
Linearity, 140
logic gates, 323
Longitudinal waves, 169
Loudness, 215
Luminescence, 181
Magnetic field, 222
Magnetic induction, 220
Magnetic relays, 229
Magnified, 201
Manometer, 73
Mass, 9, 21
Mass number, 343
mercury barometer, 72
Metre, 7
Milky Way, 102
millibar, 72
Moments, 49
momentum, 97
Motor effect, 289
NAND, 326
neutral equilibrium, 57

neutron, 342
Newton, 15, 21
Newton's First Law, 26
Newton's Second Law, 27
Newton's Third Law, 29
Non-Renewable fuels, 87
NOR, 327
normal, 171
NOT, 325
Nuclear, 75
NUCLEAR, 87
nucleus, 217
Nuclide, 343
Ohm's Law, 255
OIL, 87
Optical Fibres, 197
OR, 324
output, 309
Parallax, 13
Parallel, 250
Pascal, 64
Perspex, 234
phases, 150
Photocopiers, 242
Pitch, 215
pivot, 49
plugs, 275
plum pudding, 338
poles, 219
polythene, 233
Potential, 75
Potential Difference, 246
Potential Dividers, 309
Power, 94
power conservation equation, 301
power dissipation in a resistor, 271
prefixes, 7
prisms, 196
processor, 309
proton, 342
Radiated, 75
Radiation, 161
Radioactive Decay, 360
Range, 140
rarefactions, 169
Reaction, 15

reaction time, 37
Real, 200
Rectifiers, 311
red shift, 106
Reflection, 173, 181
Refraction, 175, 186
refractive index, 189
Regular reflection, 181
Relays, 309
repel, 224
Resistance, 254
Resistors, 309
right hand coil clasp rule, 229
Right Hand Grip Rule, 227
Right-hand Rule, 281
Ripple Tanks, 173
Sankey diagrams, 77
Scalar, 23
Scientific notation, 8
Seconds, 7
Sensitivity, 140
Series, 249
Shells, 217
Snell's law, 194
soft magnetic material, 220
Solar, 92
SOLAR, 91
Sound, 75
Specific Heat Capacity, 146
specific latent heat of boiling., 155
specific latent heat of fusion, 150
speed of sound, 214
split ring commutators, 295
Stability, 56
stable equilibrium, 56
static electricity, 233
step down transformer, 299
step up transformer, 299
step-down transformers, 305
Sterilising, 358
Stopping distances, 37
Temperature, 139
Tension, 15
The coil, 227
the national grid system, 304
the **principle of moment**, 51
thermal capacity, 146

thermionic emission, 331
thermistor, 317
Thermocouple, 140
Thinking Distance, 37
Tidal, 92
Tidal power, 88
Time, 10
time period, 168
Time Period, 167
Torsion, 15
total internal reflection, 193
Tracers, 357
transducers, 308
Transformer efficiency, 303
Transformer Equation, 300
transformers, 220
Transformers, 298
Transistors, 309, 312
transmission of power, 305
Transverse waves, 169
Troughs, 167
Truth tables, 323
Turning Forces, 49
Unbalanced force, 26
unstable, 57
Vacuum flask, 162
valleys, 167
Van De Graaf Generator, 238
Vector, 23
Velocity-time graphs, 35
Virtual, 201
Voltage, 246
Voltmeter, 246
Wave, 92
WAVE, 90
waveforms, 335
wavelength, 167
weight, 15
Weight, 16, 21
Wind, 92
Wind farms, 90
Work, 94
X-plates, 332
Y-plates, 332

Printed in Great Britain
by Amazon